Sangaku Reflections

Sangaku Reflections

A Japanese Mathematician Teaches

J. Marshall Unger

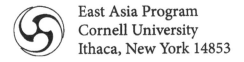

East Asia Program
Cornell University
Ithaca, New York 14853

The Cornell East Asia Series is published by the Cornell University East Asia Program (distinct from Cornell University Press). We publish books on a variety of scholarly topics relating to East Asia as a service to the academic community and the general public. Address submission inquiries to CEAS Editorial Board, East Asia Program, Cornell University, 140 Uris Hall, Ithaca, New York 14853-7601.

Cover Illustration
Front: Portrait of Aida Yasuaki mounted on a hanging scroll (kakejiku) in the collection of the Japan Academy (Nihon Gakushi-in) and reproduced with its permission and kind assistance.
Back: Utagawa Hiroshige (1797–1858). Kazusa Yazashi-ga-ura tsumei
六十余州名所図会　上総　矢さしか浦通名九十九里
Yasashi Beach, known as Kujūkuri, Kazusa Province, from the series Views of Famous Places in the Sixty-Odd Provinces Yasashi Beach, known as Kujūkuri, Kazusa Province, from the series Edo Period, ca. 1853. Polychrome woodblock print; ink and color on paper. Purchase, Joseph Pulitzer Bequest, 1918. JP596.
https://metmuseum.org/art/collection/search/37047

Number 189 in the Cornell East Asia Series
Copyright ©2017 Cornell East Asia Program. All rights reserved.
ISSN: 1050-2955
ISBN: 978-1-939161-69-7 hardcover
Library of Congress Control Number: 2018936993

In grateful memory of Martin Gardner

Contents

viii

Preface

This book is a companion to my earlier volume *Sangaku Proofs: A Japanese Mathematician at Work*. It is an interpretive translation of a one-fascicle book by Aida Yasuaki 会田安明 (1747–1817) entitled *Kajiten sandaishū*, A Collection of Calculation Problems that Require Addenda 可辞添算題集. Like *Saijō-ryū sanpō kantsū jutsu* 最上流算法貫通術, about which I wrote earlier, it is a manuscript, but differs from it in three ways.

First, while *Sanpō kantsū jutsu*, comprising more than sixty fascicles, is a kind of treatise covering a rich variety of geometric problems, one can tell from Aida's introduction to single-fascicle *Kajiten sandaishū* that its scope is narrower and its purpose more didactic. Aida's goal is to alert the reader to the need to state side conditions of problems explicitly from time to time. This is the reason for the Sino-Japanese phrase *kajiten*—or *kotoba sourubeki* as it would be glossed in Japanese—in the title. The phrase means '[those which] words must accompany'. In keeping with his purpose, Aida presents problems in pairs, showing how a slight difference in side conditions can affect how the solution is reached.

Second, in *Kajiten sandaishū*, Aida sometimes omits lengthy proofs, instead referring the reader to passages in both *Sanpō kantsū jutsu* and his massive *Saijō-ryū sanpō tenseihō*, The Saijō School Principles of Algebraic Calculation 最上流算法天生法, in more than a hundred fascicles. Such cross-referencing would not be of much help to a reader who lacked ready access to these works, which suggests that Aida probably had his own disciples in mind when he wrote *Kajiten sandaishū*. It also shows that this book must have been a late creation. As noted above, Aida died in 1817, and his *Sanpō tenseihō shinan*, a five-fascicle guide to *Sanpō tenseihō*, dates from 1810. The work to which it serves as an annotated table of contents could not have been completed much earlier given its great length. A safe guess is therefore that *Kajiten sandaishū* was written roughly between 1810 and the time that Aida's health began to fail.

Finally, like other practitioners of Japanese mathematics (*wasanka*), Aida poses problems and gives their solutions in a variety of literary Chinese called *kanbun*, but uses more or less ordinary Japanese[1] to explain points that might have been hard to comprehend if written only in *kanbun*. Even a cursory examination of both texts suggests that he does this more often in *Kajiten sandaishū* than in *Sanpō kantsū jutsu*. Although he could have been adding notes for his own reference, it seems more likely that he had students in mind. Also, as in the case of problem pairs 6 and 7, the real complexity of the solutions lies not in the point being illustrated in *Kajiten sandaishū*, which may involve nothing more than a change of a sign in an algebraic expression, but rather in the theorem, recorded elsewhere, that provides the essential tool for computing the solution. I cannot help suspecting that Aida included these problem pairs to stimulate his students' self-study.

For all these reasons, I think that *Kajiten sandaishū* was meant as teaching material—hence my choice of title—but emphasize that this conjecture is based on internal features of the text. To show readers how modern Japanese academic historians of mathematics make use of external evidence and comparison of texts when studying individual works, I include an essay in Appendix A on a better-known one-fascicle manuscript by Aida called *Oranda*

---

[1] I say "more or less" because the kind of Japanese used was in a style that facilitated the glossing of the individual characters of a *kanbun* text, transposing them as necessary, and adding conventional inflections and grammatical particles to make the result hang together in an intelligible way. This technique of rendering *kanbun*, called *yomikudashi*, reliably but mechanically produced expressions that must have sounded stilted and old-fashioned compared with contemporary conversational speech.

sanpō Dutch Calculation 阿蘭陀算法. The article about it by Kobayashi Tatsuhiko, on which I focus in the essay, sheds light on the relationship between Aida and one of his own teachers, the redoubtable Honda Toshiaki, and affords several opportunities to explain aspects of Edo period mathematics that do not come up in *Kajiten sandaishū* or the portions of *Sanpō kantsū jutsu* I translated previously.

დ႞ა

In this book, I use a more liberal method of labeling figures than I did previously, not bothering to encode the Chinese characters (*kanji*) that Aida chose the same way in every problem.

As in my previous book, I follow the order of exposition that Aida uses, and faithfully render what he wrote. But for two reasons, I make no attempt to offer a word-by-word rendition of the laconic literary Chinese (*kanbun*) in which Aida writes most of the text. First, a translation of that kind would be cryptic and defeat the aim of making the text accessible to a wide readership. (For the same reason, I make free use of modern mathematical notation and terminology as needed.) Second, Aida occasionally slips or pursues an unexpected line of reasoning. It would be pedantic to translate the text exactly only to set things right immediately thereafter. Instead, when necessary, I drop the voice of Aida as first-person narrator and explain the difficulties of the text without undue fuss. Although this somewhat blurs the line between author and translator, I have taken care to make it clear who is saying what. The sections labeled "Remarks" are my own observations except where I explicitly indicate that I am translating from primary source material.

Primary source material in the *wasan* tradition is voluminous. I have listed only modern works in the references. I have so far relied on digitized versions of works by Aida and other *wasanka* in the Yamagata University Library Sakuma Collection and on the archived materials accessible on the website of Hiroshi Kotera. The Tōhoku University Library also maintains a large, well-indexed collection of digitized *wasan* materials, which I have only just begun to explore.

I would like to thank Harald Kümmerle, who is completing a dissertation at the University of Halle on the institutionalization of higher mathematics in Meiji and Taishō Japan, for generously sharing with me his knowledge of modern Japanese research on *wasan*. He and I met at a conference at the end of April 2017 in honor of Fukagawa Hidetoshi organized by David A. Clark of Randolph-Macon College in Ashland, Virginia. My thanks also to Em. Prof. Kobayashi Tatsuhiko, with whom Kümmerle put me in touch; I have benefited greatly from his comments on the draft of Appendix A.

Finally, as this book is produced from photo-ready pages I have composed myself, let me thank the mathematical proofreaders who checked the text at the behest of my diligent editor Mai Shaikhanuar-Cota. Any errors that may remain are entirely my responsibility.

I hope the small contribution the present volume makes to the English-language literature on *wasan* helps foster more international explorations of its mathematical beauty and historical significance.

Aida begins *Kajiten sandaishū* with a preface that strongly suggests a didactic intent.

> There is a rule for stating problems for study. When you add words carelessly, a problem is spoiled; when you suppress words carelessly, it may turn into a different problem. Therefore, you must use all and only necessary words: this is the rule.
>
> But even when you state a problem and state a solution that way, there are cases in which the solution does not hold when one changes the given values. Or the problem requires reasoning different from that which works in many ordinary cases; even problems that ask for an integer n that yields remainders $r_1, r_2, \ldots, r_k$ when divided by $m_1, m_2, \ldots, m_k$, respectively, may not, depending on the values given, be soluble with the same method.[2] Each such problem is of its own kind, where we understand all problems with the same method of solution regardless of changes in the given values to be of the same kind.
>
> The rule is modified, however, for problems in which a systematic change in given values leads to a different analysis. When encountering such problems, one should accompany each case with the proper words and not digress into variations. As such problems arise only in rare situations, I offer some of them below by way of example, with explanations and solutions.

I put the crucial proper words in boldface in the problems below.

---

[2] Apparently, Aida was thinking of the Chinese remainder theorem, which is the key to solving many such problems. It requires that the divisors be pairwise coprime. If they are not, a solution may exist but must be found some other way.

Problem 1a

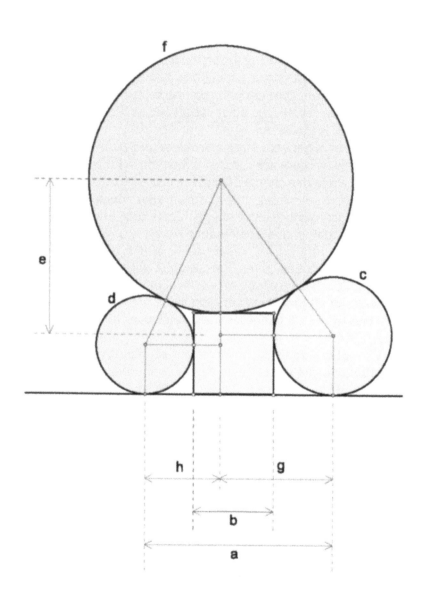

A contact chain of two circles with diameters $c > d$ and a square of side b sit on a line as shown. **The circle centers lie below the top of the square.** A third circle with diameter f touches all three links in the chain. If $c = 13, d = 10$, and $b = 9$, what is f?

ANSWER: $f = 37\frac{25}{36}$.

METHOD: From $a = b + \frac{c+d}{2}$ and $e + \frac{c}{2} = b + \frac{f}{2} \Leftrightarrow e = b - \frac{c-f}{2}$, we have

$$g^2 = \left(\frac{c+f}{2}\right)^2 - e^2 = cf + b(c - f) - b^2 = (b + f)(c - b).$$

Thus $g = \sqrt{b + f}\sqrt{c - b}$ and, likewise, $h = \sqrt{b + f}\sqrt{d - b}$. Hence

$$\sqrt{b + f}\sqrt{c - b} + \sqrt{b + f}\sqrt{d - b} - a = 0.$$

Let $k = \frac{\sqrt{c-b}}{\sqrt{d-b}} + 1$. Then $k\sqrt{b + f}\sqrt{d - b} - a = 0$. Squaring, $k^2(b + f)(d - b) - a^2 = 0$. Therefore $f = \frac{(a/k)^2}{d-b} - b$. $\square$

Problem 1b

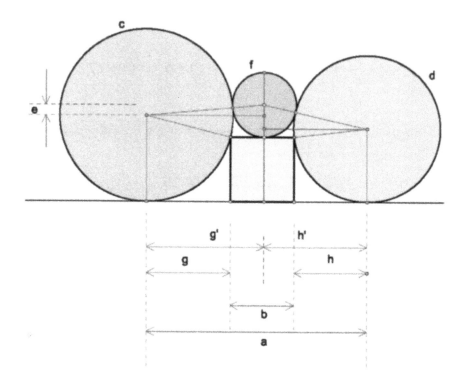

A contact chain of two circles with diameters $c > d$ and a square of side b sit on a line as shown. **The circle centers lie above the top of the square.** A third circle with diameter f touches all three links in the chain. If $c = 34, d = 25$, and $b = 9$, what is f? ANSWER: $f = 7$.

METHOD: It is easy to see that $g^2 = \left(\frac{c}{2}\right)^2 - \left(\frac{c}{2} - b\right)^2 = b(c - b)$. Likewise $h^2 = b(d - b)$. Note that $a = g + b + h = g' + h'$.

By the Pythagorean theorem, $g'^2 = \frac{(c+f)^2}{4} - e^2$. But as before $e = b - \frac{c-f}{2}$; therefore

$$g'^2 = \frac{(c + f)^2}{4} - \left(b - \frac{c - f}{2}\right)^2 = cf + b(c - f) - b^2$$
$$= (b + f)(c - b).$$

Likewise, $h'^2 = (b + f)(d - b)$. Squaring $a - h' = g'$, we get $a^2 - 2ah' + h'^2 = g'^2$ or $g'^2 - h'^2 + 2ah' - a^2 = 0$. Replacing g' and h',

$$(b + f)(c - d) + 2a\sqrt{(b + f)(d - b)} - a^2 = 0.$$

And letting $ax = \sqrt{b + f}$, this is $(c - d)x^2 + 2x\sqrt{d - b} - 1 = 0$, which has the roots $x = \frac{\pm\sqrt{c-b} - \sqrt{d-b}}{c-d}$. Only the positive root $\frac{\sqrt{c-b} - \sqrt{d-b}}{c-d}$ is relevant, and Aida therefore writes $(c - d)x - \sqrt{c - b} + \sqrt{d - b} = 0$.

It follows that $(c - d)\sqrt{b + f} - \left(a\sqrt{c - b} - a\sqrt{d - b}\right) = 0$. Because $\sqrt{c - b} - \sqrt{d - b} = \frac{g-h}{\sqrt{b}}$, this can be written as $\sqrt{b + f} = \frac{a(g-h)}{(c-d)\sqrt{b}}$, so $f = \frac{a^2(g-h)^2}{b(c-d)^2} - b$. $\square$

REMARKS ON PROBLEM PAIR 1

The reason that 1a and 1b have different solutions is obvious when we think about the centers of the circles that touch the base-line (bottom of the square extended). They lie on a straight line (with slope −1 in the figure below) for circles that touch the adjacent side of the square, but on a parabola for circles that pass through its endpoint. The focus of the parabola is the endpoint; the directrix is the baseline; the common point of the two loci is endpoint of the latus rectum of the parabola.

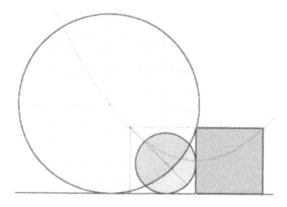

Looking at things from the viewpoint of construction, the yellow square in figure 1b corresponds to the darker square in the following figure on the left when both radii are less than or equal to the side of the square, i.e. $a = \frac{c}{2} + b + \frac{d}{2}$. But if, say, $\frac{c}{2} > b$, as shown on the right, the yellow square of 1b corresponds to the lighter colored square.

If both radii are greater than b, then each center lies on its own parabola.

In 1a, assuming that $c = d$ and $b = \frac{c}{2}$, we have $k = 2$ and $c = d = 2b$, so $a = 3b$ and $f = \frac{9b^2}{4b} - b = \frac{5b}{4}$. This cannot be obtained from the formula in 1b, which is useless when $c - d = 0$. In that case, if $c > 2b$, one must rely on the fact that $g' = g + \frac{b}{2}$, which leads to $f = \frac{b\left[b + 4\sqrt{b(c-b)}\right]}{4(c-b)}$. Aida does not mention this.

Instead, he concludes with another way to simplify $\sqrt{b+f}\sqrt{c-b} + \sqrt{b+f}\sqrt{d-b} - a = 0$ in solution 1a. Using the terms indicated in the figure, this equation is $\frac{(g+h)\sqrt{b+f}}{\sqrt{b}} - (g' + h') = 0$. Unfortunately, he miswrites this as $\frac{(g+h)\sqrt{b+f}}{\sqrt{b}} - (g + h) = 0$ (which implies $f = 0$), failing to take into account that g, h in 1a are g', h' in 1b.

Aida's oversights in his discussion of these problems (and elsewhere) tend to suggest that he was a confident, compulsive, and rather impatient writer. Because much of the *wasan* literature, for multiple reasons, circulated in the form of hand-copied manuscripts (Nishida 2013), Aida probably prepared manuscripts for publication by woodblock printing only on rare occasions. Composing a new manuscript, which could contain new ideas as well as corrections, was almost as easy as rereading and correcting the older one it superseded, and was perhaps a more enjoyable task.

Problem 2a

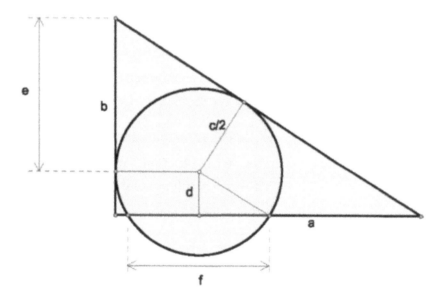

Given a right triangle, a circle of diameter c **centered inside the triangle** contacts the hypotenuse and the leg measuring b as shown. If $a = 5.2, b = 3.9$, and $c = 3$, what is the length of the chord f on leg a? ANSWER: $f = 2.4$. Furthermore, the hypotenuse of the triangle is 6.5, the diameter of the incircle is $g = 2.6$, and $d = 0.9$.

METHOD: Because we have a right triangle, the diameter of the incircle g is $a + b - \sqrt{a^2 + b^2}$, and by similar triangles, $\dfrac{b - \frac{g}{2}}{g} = \dfrac{e}{c}$.

Therefore $d = b - e = b - \dfrac{c\left(b - \frac{g}{2}\right)}{g}$. From the figure, we see that

$\left(\dfrac{f}{2}\right)^2 = \left(\dfrac{c}{2} + d\right)\left(\dfrac{c}{2} - d\right)$ or $\dfrac{f^2}{4} = \dfrac{c^2}{4} - d^2$, which Aida carelessly

writes as $f^2 = \dfrac{c^2}{4} - d^2$ instead of $f^2 = c^2 - 4d^2$. But he immedi-

ately gives the formula $f = \sqrt{c^2 - \left[2b - \dfrac{c(2b-g)}{g}\right]^2}$, which is equiv-

alent to $f^2 = c^2 - 4\left[b - \dfrac{c(b-g/2)}{g}\right]^2$, correcting the error. $\square$

Problem 2b

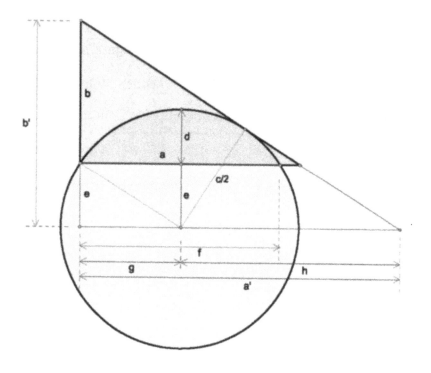

Given a right triangle, a circle of diameter c **centered outside the triangle** contacts the hypotenuse and the leg measuring b as shown. If $a = 4, b = 3$, and $c = 7.8$, what is the length of the chord f on leg a? ANSWER: $f = 3.96$. Furthermore, the sagitta d is 0.54.[3]

METHOD: In the figure, $g = \frac{f}{2}, \frac{c}{2} - d = e$, $b + e = b'$, and, since $e^2 + g^2 = \frac{c^2}{4}$, we have

$$f^2 - 4cd + 4d^2 = 0.$$

By similar triangles, $a' = \frac{ab'}{b}$ and $h = \frac{c\sqrt{a^2+b^2}}{2b}$. Hence the equation $g + h - a' = 0$ may be rewritten $2bg + c\sqrt{a^2 + b^2} - 2ab' = 0$. Replacing b' with $b + \frac{c}{2} - d$, this is $2bg + c\sqrt{a^2 + b^2} - 2a\left(b + \frac{c}{2} - d\right) = 0$ or, letting $j = \sqrt{a^2 + b^2}$ and rearranging,

$$c(j - a) + bf - 2ab + 2ad = 0.$$

Substituting the solution of the last equation for d into the first, we get a quadratic in f, which Aida writes initially as

$$4a^2b^2 - 4a^2bc - 4ab^2f + 2abcf + a^2f^2 + b^2f^2 - 4abc(j - a)$$
$$+ 2ac^2(j - a) + 2bcf(j - a) + c^2(j - a)^2 = 0.$$

He simplifies this in three steps (with some errors in writing) starting with

---

[3] A SAGITTA (Latin 'arrow') is the line segment joining the midpoints of the chord and arc of a circular segment. Think of the arc as a bow and the chord as its bowstring.

$$4a^2b^2 + b^2c^2 - 4ab^2f - 4abcj + 2bcfj + j^2f^2 = 0.$$

Then, defining $k = cj - ab$,

$$-4abk + b^2c^2 + 2b(k - ab)f + j^2f^2 = 0.$$

Aida writes the discriminant as $4a^3bk$, but it is actually $16a^3bk$. Thus we have roots $\frac{b(ab-k)\pm 2a\sqrt{abk}}{j^2}$, of which $\frac{b(ab-k)+2a\sqrt{abk}}{j^2}$ is the one relevant to the problem. $\square$

REMARKS ON PROBLEM PAIR 2

Again, although Aida wants the two problems to appear identical except for the boldfaced provisions, the locus of the circle centers in 2b is the parabola for which the directrix is the hypotenuse and the focus is the right vertex, whereas, in 2a, it is the bisector of the angle opposite the side that the circle meets a second time.

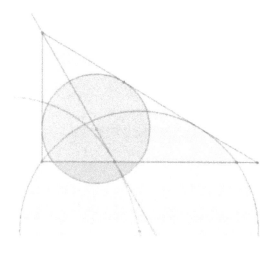

In 2a, it is obvious both algebraically and geometrically that $f = c$ for $d = 0$. This cannot be established algebraically from the formula in 2b, from which d has been eliminated. It is also worth noting that the Japanese typically referred to diameters rather than radii, and multiplied expressons through by 2/2 to get rid of pesky factors of ½. Perhaps this contributed to Aida's miswriting the equation for f^2 before giving the correct narrative recipe.

In 2b, instead of stating $f = \frac{b(ab-k)+2a\sqrt{abk}}{j^2}$, Aida writes $j^2f - ab^2 + bk - 2a\sqrt{abk} = 0$ and the equivalent equation $j^2f - aj^2 + a^3 + bk - 2a\sqrt{abk} = 0$. Based on these, he states the solution as

$$f = \frac{-bk + a(b^2 + 2\sqrt{abk})}{j^2}$$

and as

$$f = a - \frac{a\left(\sqrt{\dfrac{bk}{a}} - a\right)^2}{j^2},$$

respectively.

Problem 3a

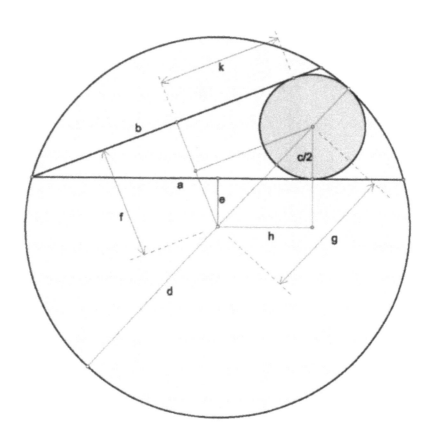

Chords a and b share a common point on the circumference of circle of diameter d. The mixtilinear triangle formed by these chords and the arc subtended by the angle they form has an incircle of diameter c.[4] **The center of the larger circle lies outside this triangle.** Given $d = 25, a = 24$, and $c = 5$, what is b? ANSWER: $b = 21\frac{7}{13}$.

METHOD: It is readily apparent in the figure that

$$e^2 = \frac{d^2}{4} - \frac{a^2}{4}, \qquad f^2 = \frac{d^2}{4} - \frac{b^2}{4}, \qquad g = \frac{d}{2} - \frac{c}{2},$$

$$h^2 = g^2 - \left(\frac{c}{2} + e\right)^2, \qquad k = \frac{a}{2} + h - \frac{b}{2},$$

and $k^2 + \left(f - \frac{c}{2}\right)^2 - g^2 = 0$. Substituting $\frac{a}{2} + h - \frac{b}{2}$ for k and $h^2 + \left(\frac{c}{2} + e\right)^2$ for g^2 on the left side of the last equation, we expand and then substitute $\frac{d^2}{4} - \frac{a^2}{4}$ and $\frac{d^2}{4} - \frac{b^2}{4}$ for e^2 and f^2, respectively, in the result. This yields

$$h(a - b) + \frac{a(a - b)}{2} - c(e + f) = 0.$$

---

[4] A MIXTILINEAR TRIANGLE is a triangle in which one side has been replaced by the corresponding arc of its circumcircle (Bankoff 1983).

Hence

$$cf = \left(h + \frac{a}{2}\right)(a - b) - ce$$

or, squaring,

$$c^2e^2 - c^2f^2 - 2ce\left(h + \frac{a}{2}\right)(a - b) + \left(h + \frac{a}{2}\right)^2(a - b)^2 = 0.$$

Replacing e^2 and f^2 in this equation,

$$-\frac{a^2c^2}{4} + \frac{b^2c^2}{4} - 2ce\left(h + \frac{a}{2}\right)(a - b) + \left(h + \frac{a}{2}\right)^2(a - b)^2 = 0$$

or

$$-\frac{(a^2 - b^2)c^2}{4} - 2ce\left(h + \frac{a}{2}\right)(a - b) + \left(h + \frac{a}{2}\right)^2(a - b)^2 = 0.$$

Multiplying through by $\frac{4}{a-b}$,

$$-(a + b)c^2 - 4ce(2h + a) + (2h + a)^2(a - b) = 0,$$

which, as a linear equation in b, is

$$-ac^2 - 4ce(a + 2h) + a(a + 2h)^2 - [c^2 + (a + 2h)^2]b = 0.$$

Now,

$$h^2 = g^2 - \left(e + \frac{c}{2}\right)^2 = \frac{(d-c)^2}{4} - \frac{d^2 - a^2}{4} - ce - \frac{c^2}{4}$$

$$= -\frac{cd}{2} + \frac{a^2}{4} - ce,$$

so $2h = \sqrt{a^2 - 2cd - 4ce}$. But $e = \frac{1}{2}\sqrt{d^2 - a^2}$, so if we let $E = 2e$ and $H = 2h$, $E = \sqrt{d^2 - a^2}$, $H = \sqrt{a^2 - 2cd - 2cE}$, and the equation for b may be written

$$-ac^2 - 2cE(a + H) + a(a + H)^2 - [c^2 + (a + H)^2]b = 0.$$

(In the text, Aida uses the same Chinese characters for e, h and E, H, so this simplification is less than transparent.) Substituting 25, 24, 5 for d, a, c, respectively, we get $35000 - 1625b = 0$, so $b = 21\frac{7}{13}$. $\square$

Problem 3b

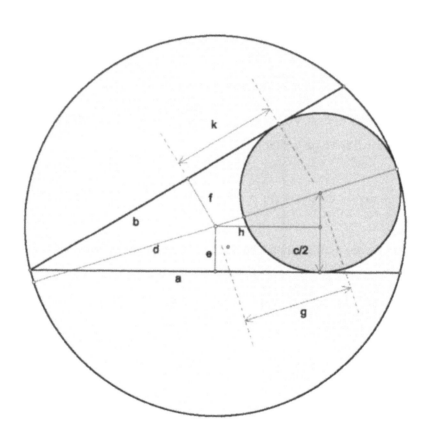

Chords a and b share a common point on the circumference of circle of diameter d. The mixtilinear triangle formed by these chords and the arc subtended by the angle they form has an incircle of diameter c. **The center of the larger circle lies inside this triangle.** Given $d = 25, a = 24$, and $c = 12$, what is b? ANSWER: $b = 23.4$.

METHOD: The basic equations for 3a hold except for a crucial change of sign: $h^2 = g^2 - \left(\frac{c}{2} - e\right)^2$.[5] Running through the same algebra as before, we get

$$-ac^2 + 4ce(a + 2h) + a(a + 2h)^2 - [c^2 + (a + 2h)^2]b = 0$$

with

$$2h = \sqrt{a^2 - 2cd + 4ce}.$$

Introducing $E = 2e$ and $H = 2h$ as before, and substituting 25, 24, 12 for d, a, c, we get $33696 - 1440b = 0$, so $b = 23.4$. $\square$

_____

[5] Since $\left(f - \frac{c}{2}\right)^2 = \left(\frac{c}{2} - f\right)^2$, we need not alter the equation for k^2.

REMARKS ON PROBLEM PAIR 3

Aida does not give the foregoing proofs in *Kajiten sandaishū*, but rather directs the reader to fascicle 31 of his voluminous *Sanpō tenseihō*.[6] Before giving the proofs there, Aida explains that this is Problem 8 in the book *Sangaku shōsen*, A Small Collection of Studies in Calculation, and spends several pages criticizing its author's original and revised solutions. Aida had already briefly discussed their defects in a critical edition of *Sangaku shōsen* that he evidently prepared for the use of his students.[7] We will examine the relevant part of *Sangaku shōsen* in a moment.

In *Kajiten sandaishū*, Aida gives only the final method of calculation. In both cases, $E^2 = d^2 - a^2$. Then, in 3a, he has

$$H^2 = a^2 - 2cd - 2cE$$
$$ac^2 + 2cE(a + H) - a(a + H)^2 + [c^2 + (a + H)^2]b = 0$$

and, in 3b,

$$H^2 = a^2 - 2cd + 2cE$$
$$ac^2 - 2cE(a + H) - a(a + H)^2 + [c^2 + (a + H)^2]b = 0.$$

Note that the terms containing the factor $2cE$ switch signs. This is the reason for including this problem pair in *Kajiten sandaishū*.

I have distinguished e, h from E, H for clarity, but Aida uses the same characters for e, E and for h, H, which at first makes it

---

[6] This turns out to be fascicle 30 of the manuscript in the Sakuma Collection at Yamagata University.
[7] The cover title of this edition is *Sangaku shōsen hyōrin* (A Treasury of Criticisms of ...). The preface of the edition, however, gives the title as *Sangaku shōsen shinjutsu* (True Methods for ...).

look as if he has made a mistake. Actually, he has just made an algebraic simplification, in effect scaling up the geometric figures by a factor of 2.

The preface of Aida's edition of *Sangaku shōsen* reads as follows:

The book *Sangaku shōsen* was written by Ushijima Ueita Moritsune of the Hosokawa domain in Kumamoto, Higo province, in 1793.[8] Nine years earlier, I had posted a *sangaku* at Yushima shrine in Edo with two self-asked and self-answered problems and five challenge problems. A disciple of mine named Watanabe posted solutions for all seven problems; compared to these, the two solutions of Moritsune were each circuitous or off-point. Now, seeing that Ushijima's book has more refined expositions, I realize how much progress he has made as he has grown older, and feel certain that he will go on to become highly accomplshed in calculations in coming years.

Nonetheless, of the 53 problems in *Sangaku shōsen*, 15 fail to state numerical values; or were copied from earlier books without acknowledgment; or from acknowledged books the titles of which are not given; or contain other kinds of omissions; and so on. Knowing that many people who cultivate the study of calculation enjoy problems involving integers most of all, I cannot help having feelings of dissatisfaction with this book.

Moreover, it contains many instances of alternative solutions of problems solved in older books. First, there are two from the works of Seki [Takakazu], both of which are circuitous. Then there are two from *Senbi sanpō*,[9] the first of which is elegant and the other fairly good. Next

---

[8] It was published in 1794.
[9] Takeda Saibi, 1750.

come five from *Shūki sanpō*,[10] of which three are circuitous. Likewise, there are five from *Seiyō sanpō*, [11] of which three are circuitous. Then one from *Shinpeki sanpō*,[12] again circuitous. Though Moritsune may have come up with different solutions, they are generally roundabout.

Furthermore, even among his self-asked and self-answered problems, a great many are similarly flawed. In particular, in number 2, the answer is too large; in 8, the chord is too large by a factor of 4; in 24, another chord is too large; and in 26, the diameter of circle A is too large. Though these are small errors, they can only be abhorred by those trying to study calculation nowadays. Finally, there are a lot of poorly stated problems. In short, *Sangaku shōsen* has many defects.

Therefore I call this book, in which I append the corrections of these defects, *True Methods for* Sangaku shōsen. As it is a book for the benefit of the young, if disciples of other schools read it, may they not be harshly punished.

1794, 3rd month

Saijō school, Aida Sanzaemon Yasuaki, Editor

Recall that our Problem 3a is Problem 8 in Ushijima's *Sangaku shōsen*. Without proving his own solution in his edition of that work, Aida zeroes in on two features of Ushijima's original proof. First, Aida notes, it takes 77 *kanji* (Chinese characters) to write down and includes three placeholders (labels for intermediate results invoked later in the recipe). Aida shows this is prolix by giving an equivalent statement in 58 *kanji* and just one placeholder. But, not stopping there, he criticizes Ushijima for, in effect, scaling

---

[10] Arima Yoriyuki, 1762.
[11] Fujita Sadasuke, 1781.
[12] Fujita Sadasuke, 1789.

up the figure by 4 rather than by 2, as does Aida. Repairing this error, Aida gives a solution in only 61 *kanji* with two placeholders, a slightly more compact version of the one he gave later in *Kajiten sandaishū*:

$$E = \sqrt{d^2 - a^2}$$
$$H' = a + \sqrt{a^2 - 2cd - 2cE}$$
$$b = \frac{2cEH' + a(c^2 - H'^2)}{c^2 + H'^2}.$$

The article in *Sanpō tenseihō* is an updated and expanded version of the foregoing criticisms, this time including the distinction between cases 3a and 3b, which is the reason for Aida's mention of of Ushijima's problem a third time in *Kajiten sandaishū*.

Given all the comparing of texts and copying Aida must have done when returning to this problem on two occasions, it is perhaps understandable that both the *Kajiten sandaishū* and *Sanpō tenseihō* texts have lapses. For instance, in the discussion of 3a in *Sanpō tenseihō*, $4h^2 = a^2 - 2cd - 4ce$ is unaccountably miswritten as $4h^2 = a^2 - 2cd - 3ce$; the metrical solution is given as 35000/2625 instead of 35000/1625; and, in the discussion of 3b, 百 '100' is miswritten in one place as 万 '10,000'.

Finally, consider the case of $a = d$, which Aida does not mention. In terms of the foregoing analysis, this means $e = 0$. Since $2h = \sqrt{a^2 - 2cd \pm 4ce}$, with the sign depending on whether we start with 3a or 3b, $2h = \sqrt{a^2 - 2cd}$.

Problem 4a

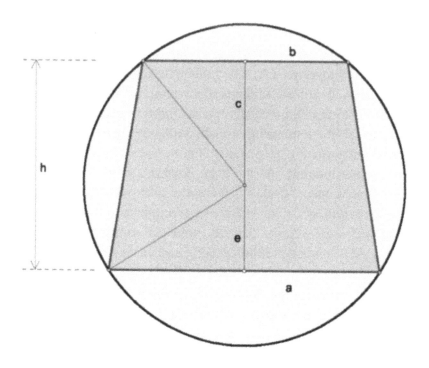

A trapezium[13] with bases $a > b$ is inscribed in a circle of diameter d, **the center of which lies within the trapezium.** If the bases are 52 and 25, and the diameter of the circumcircle is 65, what is the altitude h of the trapezium? ANSWER: $h = 49.5$.

METHOD: From $c^2 = \frac{d^2}{4} - \frac{b^2}{4}$ and $e^2 = \frac{d^2}{4} - \frac{a^2}{4}$, we immediately have $2c = \sqrt{d^2 - b^2}$ and $2e = \sqrt{d^2 - a^2}$.

Therefore $h = \frac{1}{2}\left(\sqrt{d^2 - b^2} + \sqrt{d^2 - a^2}\right)$. □

---

[13] Americans call a quadrilateral with one pair of parallel sides a TRAPEZOID, but TRAPEZIUM is used elsewhere in English, in Late Latin, and so too in French (*trapèze*), etc. Since the source, Greek *trápeza* < older *trapézion*, means 'table' and the *-oid* suffix 'likeness' (Greek again) indicates similar but not identical form, trapezium is the better choice etymologically.

Problem 4b

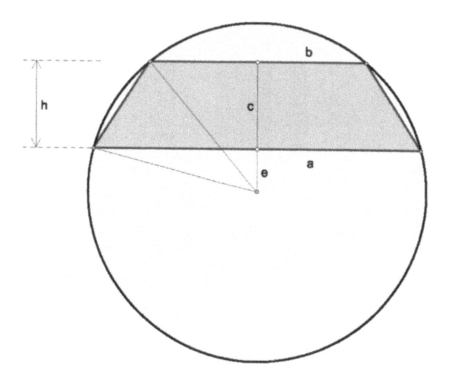

A trapezium with bases $a > b$ is inscribed in a circle of diameter d, **the center of which lies outside the trapezium.** If the bases are 52 and 25, and the diameter of the circumcircle is 65, what is the altitude h of the trapezium? ANSWER: $h = 10.5$.

METHOD: This time we have $h = \frac{1}{2}\left(\sqrt{d^2 - b^2} - \sqrt{d^2 - a^2}\right)$ because $h = c + e$ instead of $c - e$ as in 4a. □

REMARKS ON PROBLEM PAIR 4

This is perhaps the most basic illustration of the kind of problem that Aida sought to address in *Kajiten sandaishū*. The fact that he included it underscores the importance he attached to obtaining metrical solutions. No wonder he saw the omission of numerical values as a defect in Ushijima's *Sangaku shōsen*!

Problem 5a

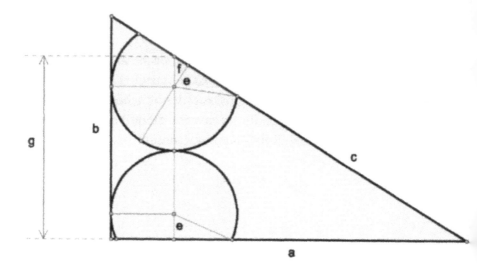

Two circles of diameter d touch externally. **Their centers lie within a right triangle**, one leg of which meets each of them. The perpendicular distance e from the center of each circle to the nearest side of the triangle is the same. If $a = 4, b = 3$, and $d = 2.1$, what is the distance from the point where the circles touch each other to leg a? ANSWER: $d/2 + e = 1.1$.

METHOD: From $\frac{a}{c} = \frac{e}{f}$, $g = d + e + f$, and $\frac{b}{a} = \frac{g}{a-d/2}$, we have

$$f = \frac{ce}{a}, \qquad g = d + e + \frac{ce}{a}$$

and

$$b\left(a - \frac{d}{2}\right) = a\left(d + e + \frac{ce}{a}\right),$$

so $ab - \frac{bd}{2} - ad - a\left(e + \frac{ce}{a}\right) = 0$, or, multiplying through by -2,
$-2ab + bd + 2ad + 2e(a + c) = 0$. We seek $x = e + \frac{d}{2}$, in terms of which

$$-2ab + bd + 2ad + 2\left(x - \frac{d}{2}\right)(a + c) = 0$$

or

$$-2ab + bd + 2ad + 2x(a + c) - d(a + c) = 0.$$

This can be shortened because $-2ab + 2x(a + c) = -ad - bd + cd = -d(a + b - c) = -dd'$, where d' is the diameter of the incircle of the triangle. That is, $2x(a + c) - 2ab + dd' = 0$. $\square$

Problem 5b

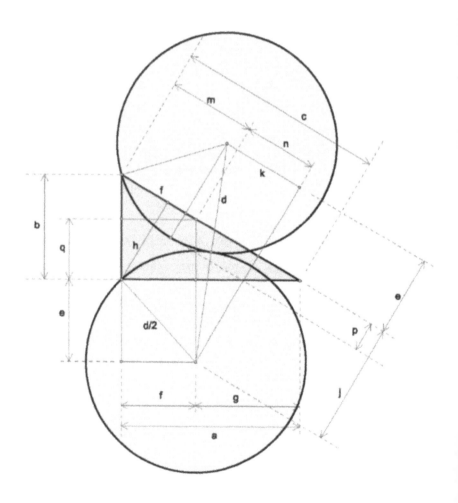

Two circles of diameter d touch externally. **Their centers lie outside a right triangle**, one leg of which meets them both. The perpendicular distance e from the center of each circle to the nearest side of the triangle is the same. If $a = 63, b = 16$, and $2f = 61$ what is the diameter of the two circles? ANSWER: $d = 317$ and a fraction.

METHOD: Aida refers the reader to fascicle 25 of *Sanpō tenseihō*, where the problem is posed without specific numerical values. In the figure, we have $g = a - f$, $q = \frac{bg}{a}$, $j = \frac{a(e+q)}{c}$, $n = \frac{b(e+q)}{c}$, $m = \frac{cf}{a}$, $k = m + n - f$, and $k^2 + (e + j)^2 - d^2 = 0$. Because $k = \frac{f(c-a)}{a} + \frac{b(e+q)}{c}$, the last equation may be rewritten as

$$\left[\frac{f(c-a)}{a} + \frac{b(e+q)}{c}\right]^2 + \left[e + \frac{a(e+q)}{c}\right]^2 - d^2 = 0$$

or

$$\frac{f^2(c-a)^2}{a^2} + \frac{2bf(c-a)(e+q)}{ac} + \frac{b^2(e+q)^2}{c^2} + e^2$$
$$+ \frac{2ae(e+q)}{c} + \frac{a^2(e+q)^2}{c^2} - d^2 = 0,$$

which, since $a^2 + b^2 = c^2$, simplifies to

$$\frac{f^2(c-a)^2}{a^2} + \frac{2bf(c-a)(e+q)}{ac} + \frac{2ae(e+q)}{c} + e^2 + (e+q)^2$$
$$- d^2 = 0.$$

Treating $e + q$ as $(e + b) - \frac{bf}{a}$, we have

$$(e + q)^2 = (e + b)^2 - \frac{2(e+b)bf}{a} + \frac{b^2 f^2}{a^2},$$

so we can eliminate q to get

$$\frac{f^2(c - a)^2}{a^2} + \frac{2bf(c - a)(e + b)}{ac} - \frac{2b^2 f^2(c - a)}{a^2 c} + \frac{2ae(e + b)}{c}$$
$$- \frac{2abef}{ac} + 2e^2 + 2be - \frac{2bef}{a} + b^2 - \frac{2b^2 f}{a}$$
$$+ \frac{b^2 f^2}{a^2} - d^2 = 0.$$

The sum of the 1st and 11th terms is

$$\frac{f^2(c - a)^2}{a^2} + \frac{b^2 f^2}{a^2} = \frac{f^2[b^2 + (c - a)^2]}{a^2} = \frac{2cf^2(c - a)}{a^2}$$

and the sum of the 2nd, 8th, and 10th terms is

$$\frac{2bf(c - a)(e + b)}{ac} - \frac{2bcef}{ac} - \frac{2b^2 cf}{ac}$$
$$= \frac{2bf[(c - a)(e + b) - ce - bc]}{ac}$$
$$= \frac{-2abf(e + b)}{ac}.$$

Moreover, $4e^2 + 4f^2 = d^2$ (Pythagorean theorem). Thus the equation simplifies to

$$\frac{2cf^2(c - a)}{a^2} - \frac{2abf(e + b)}{ac} - \frac{2b^2 f^2(c - a)}{a^2 c} + \frac{2ae(e + b)}{c}$$
$$- \frac{2bef}{c} - 2e^2 + 2be + b^2 - 4f^2 = 0.$$

The sum of the 1st, 3rd, and last terms now is

$$\frac{2c^2f^2(c-a)}{a^2c} - \frac{2b^2f^2(c-a)}{a^2c} - \frac{4a^2cf^2}{a^2c}$$

$$= \frac{2f^2[c^2(c-a) - b^2(c-a) - 2a^2c]}{a^2c}$$

$$= \frac{2f^2[a^2(c-a) - 2a^2c]}{a^2c} = \frac{-2f^2(c+a)}{c},$$

so $-\frac{2f^2(c+a)}{c} - \frac{2bf(e+b)}{c} + \frac{2ae(e+b)}{c} - \frac{2bef}{c} - 2e^2 + 2be + b^2 = 0.$

Combining the 2nd and 4th terms in this equation, multiplying through by c, and rearranging, we have

$$-2f^2(c+a) - 2b^2f + b^2c - 4bef + 2be(c+a) - 2e^2(c-a)$$
$$= 0.$$

As a quadratic in e, this is

$$b^2c - 2f^2(c+a) - 2b^2f + 2b(c+a-2f)e - 2(c-a)e^2 = 0.$$

Letting $r^2 = 4c(c-2f) - b^2$, the discriminant is $4b^2r^2$ and the roots are

$$e = \frac{-2b(c+a-2f) \pm 2br}{-4(c-a)} = \frac{b(c+a-2f) \pm br}{2(c-a)}.$$

Aida identifies the relevant root by writing the linear equation $b(c+a-2f) - br - 2(c-a)e = 0$. Then, using $4e^2 + 4f^2 = d^2$ once again, he obtains

$$d^2 = \left(\frac{b[(c+a-2f) - \sqrt{4c(c-2f) - b^2}]}{c-a} \right)^2 + 4f^2. \quad \square$$

REMARKS ON PROBLEM PAIR 5

Once again, as in pair 1, the relationship between these problems is rather tenuous from a modern perspective. And unlike previous problems, what is asked for is not the same.

In 5a, c and d' must be calculated from a and b, so Aida's mention of the diameter of the incircle is extraneous.

Aida labels segments h and p in the principal figure for 5b, but doesn't make use of them in the solution. In an introductory sketch, he gives a label to the segment $c - 2f$, which he uses only after simplifying the discriminant—as it adds nothing to understanding the algebra, I have omitted it. The reason for these quirks of notation becomes clear when one looks at the two related problems that immediately follow 5b in *Sanpō tenseihō*. The first asks for d given a, b, and $2f$ (which is now given the same label as $c - 2f$ in the previous problem); the second asks for p given a, b, and d. Aida evidently thought highly of this problem. After presenting it, he praises Gunji Fusayori Tsunenao 軍司総因陽尚 of the Mito domain, who proposed it in 1752. He explains how he obtained a copy of the problem, and concludes with the following anecdote:

> They say Fusayori was an itinerant teacher for eleven years, starting off with a single pack-horse, a manservant, and 200 *ryō* for expenses, and that he returned eleven years later, highly successful, with 800 *ryō*.[14]

---

[14] The readings of Gunji's personal names are my best guess.

Gunji was an itinerant or *yūreki* 遊歴 scholar (Rubinger 1982: 25). The fact that such people, traveling outside their domains and lecturing on *wasan*, could be financially successful shows that there was demand for instruction and that was more to *wasan* than the competitive puzzle-solving reflected in *sangaku*.

A *ryō* was an oblong gold coin, the value of which fluctuated considerably throughout the Edo period. At the time in question, the annual salary

None of this is repeated in *Kajiten sandaishū*.

Aida had been adopted into a *samurai* family in his native Yamagata before setting off to begin his career in Edo, and this anecdote hints at his non-*samurai* origins. [15] Children of *samurai* were taught that talk of money was beneath the dignity of the family.

---

of a typical laborer was about 6 *ryō*; a full formal suit of clothes for a *samurai*, about 1 *ryō*; annual tuition for a child in a temple school, about ¼ *ryō* (http://wiki.samurai-archives.com/index.php?title=Currency).

[15] A sword can been seen on a stand by the side of the famous mathematician Seki Takakazu in a well-known portrait of him, but no swords are visible in Aida's portrait.

Problem 6a

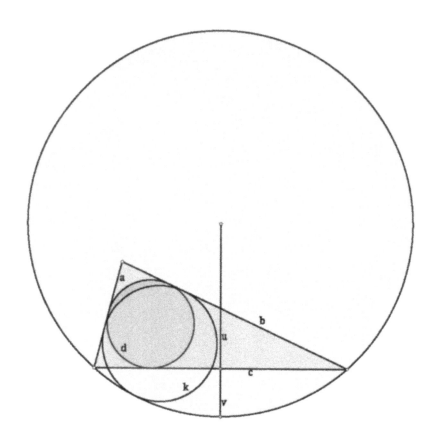

A triangle has sides $a < b < c$; c is a chord of a circle with diameter D, **the center of which lies outside the triangle,** and h is the altitude on it. The incircle of the triangle has diameter d; k is the diameter of the pseudo-incircle.[16] If $a = 8, b = 26, c = 30$, and $D = 50$, what is k? ANSWER: $k = 9$; moreover, sagitta $v = 5$.

METHOD: The solution is based on a much-discussed theorem, which Aida proved in fascicle 4 of *Sanpō tenseihō* (see the Remarks on p. 40 ff.). Note first that $u^2 = \frac{D^2}{4} - \frac{c^2}{4}$ (Pythagorean theorem) and that $v = \frac{D}{2} - u$.

Let $t_0 = b + a - c$. Then $k = \frac{2d^2v}{ct_0} + d$ or $k = \frac{4dhv}{t_0(a+b+c)} + d$.

Aida writes the recipe for k as $\left(\frac{2h(D-\sqrt{D^2-c^2})}{(b+a)^2-c^2} + 1 \right) d$. □

---

[16] Given a triangle and a circle, a PSEUDO-INCIRCLE is a circle that is tangent to two sides of the triangle and internally tangent to the circle (Rabinowitz 2006). If the given circle is the circumcircle of the triangle, then the pseudo-incircle is the incircle of a mixtilinear triangle.

Problem 6b

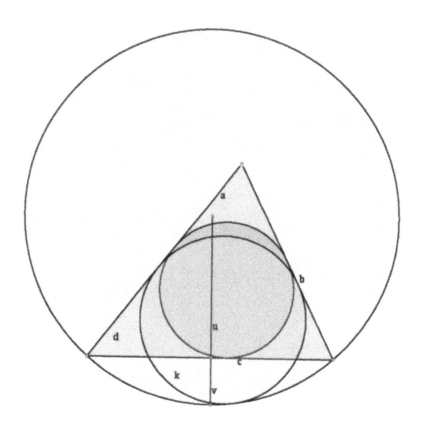

A triangle has sides $a < b < c$; c is a chord of a circle with diameter D, **the center of which lies inside the triangle,** and h is the altitude on it. The incircle of the triangle has diameter d; k is the diameter of the pseudo-incircle. If $a = 8, b = 26, c = 30$, and $D = 50$, what is k? ANSWER: $k = 9$; moreover, sagitta $v = 5$.

METHOD: The solution is the same as for 6a. The position of the center in relation to the triangle is immaterial. □

REMARKS ON PROBLEM PAIR 6

This pair of problems certainly fit Aida's stated purpose in composing *Kajiten sandaishū*, but they assume a theorem that is far from obvious and, in fact, quite a chore to prove.   It is $k = d + \frac{d^2 v}{ct}$ in reference to the figure below.

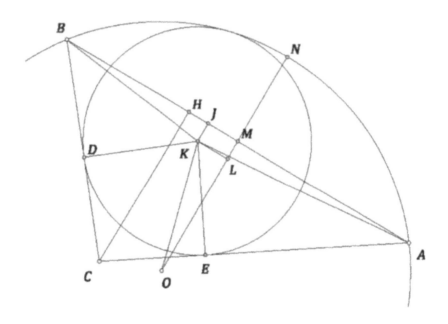

$$BC = a, AC = b, AB = c, MN = v,$$
$$CH = h, OK = l, JK = LM = m$$
$$JM = KL = n, OL = q$$
$$CD = CE = t, OM = u$$

$$NO = \frac{D}{2}, KD = KE = \frac{k}{2}, \text{inradius of } \triangle ABC = \frac{d}{2}.$$

N.B. D, d, k are diameters.

Here is Aida's proof in *Sanpō tenseihō.*

Let

$$b + a - c = t_0 \; (= 2t)$$
$$a + c - b = t_1$$
$$b + c - a = t_2$$
$$a + b + c = p$$

Using these four definitions, Aida states that

$$pt_0 = (b + a)^2 - c^2,$$
$$t_1 t_2 = -(b - a)^2 + c^2,$$
$$t_0 t_1 t_2 = d^2 p,$$
$$\frac{hc}{2} = \frac{dp}{4} = |\triangle ABC|.$$

Today, we would more likely write $r = \frac{d}{2}$. Also, since $|\triangle ABC| = \sqrt{s(s-a)(s-b)(s-c)}$ (Heron's area formula), where semi-perimeter $s = \frac{a+b+c}{2}$, we would write $\frac{b+a-c}{2} = s - c$, $\frac{a+c-b}{2} = s - b$, and $\frac{b+c-a}{2} = s - a$ rather than Aida's p, t_0, t_1, and t_2. His equations in modern form would be

$$4s(s - c) = (b + a)^2 - c^2,$$
$$4(s - b)(s - a) = -(b - a)^2 + c^2,$$
$$(s - a)(s - b)(s - c) = r^2 s,$$
$$hc/2 = rs = |\triangle ABC|.$$

The reason for mentioning these equivalences will become clear later on.

By the Crossed Chords theorem, $v + \frac{c^2}{4v} = D$. With this and the first four equations, Aida transforms the Pythagorean relation for the sides n, q, l of right triangle KLO into a formula for k in terms of $a, b, c,$ and v as follows.

First, since $m^2 = BK^2 - \left(\frac{c}{2} - n\right)^2 = AK^2 - \left(\frac{c}{2} + n\right)^2$, we have $\frac{AK^2}{2c} - \frac{BK^2}{2c} = n$. But $(b-t)^2 + \left(\frac{k}{2}\right)^2 = AK^2$, and $(a-t)^2 + \left(\frac{k}{2}\right)^2 = BK^2$. Therefore $\frac{(b-t)^2}{2c} - \frac{(a-t)^2}{2c} = n$. That is, $\frac{(b-a)(b+a)}{2c} - \frac{2t(b-a)}{2c} = n$. But by similar triangles (omitted in the figure), $\frac{d}{k} = \frac{t_0/2}{t} \Leftrightarrow t = \frac{kt_0}{2d}$. Using this to eliminate t in the foregoing equation, we have $\frac{(b-a)(b+a)}{2c} - \frac{kt_0(b-a)}{2cd} = n$.

Next, it is easy to see that $\frac{D}{2} - v = u$, and $u - m = q$. Since $\triangle ABC - \triangle AKC - \triangle BKC = \triangle AKB$, we have $hc - \frac{bk}{2} - \frac{ak}{2} = mc$, so $h - \frac{(b+a)k}{2c} = m$. Therefore $\frac{D}{2} - v - h + \frac{k(b+a)}{2c} = \frac{D}{2} - v - \frac{dp}{2c} + \frac{k(b+a)}{2c} = q$.

Finally, $\frac{D}{2} - \frac{k}{2} = l$.

To get rid of denominators in the equations for n, q, l, multiply by $2cd$:

$$d(b-a)(b+a) - kt_0(b-a) = n'$$
$$cdD - 2cdv - d^2p + dk(b+a) = q'$$
$$cdD - cdk = l'$$

Squaring each,

$$k^2 t_0^2 (b-a)^2 - 2dkt_0(b-a)^2(b+a) + d^2(b-a)^2(b+a)^2 = n'^2$$

$$c^2d^2D^2 - 4c^2d^2Dv - 2cd^3Dp + 2cd^2Dk(b+a) + 4c^2d^2v^2$$
$$+ 4cd^3pv - 4cd^2kv(b+a) + d^4p^2$$
$$- 2d^3kp(b+a) + d^2k^2(b+a)^2 = q'^2$$

$$c^2 d^2 D^2 - 2c^2 D d^2 k + c^2 d^2 k^2 = l'^2$$

so, since $n'^2 + q'^2 - l'^2 = 0$ (dilating by $2cd$ does not matter), we have

$$
\begin{aligned}
k^2 t_0^2 (b-a)^2 &- 2dk t_0 (b-a)^2 (b+a) + d^2 (b-a)^2 (b+a)^2 \\
&+ c^2 d^2 D^2 - 4c^2 d^2 Dv - 2cd^3 Dp \\
&+ 2cd^2 Dk(b+a) + 4c^2 d^2 v^2 + 4cd^3 pv \\
&- 4cd^2 kv(b+a) + d^4 p^2 - 2d^3 kp(b+a) \\
&+ d^2 k^2 (b+a)^2 - c^2 d^2 D^2 + 2c^2 D d^2 k - c^2 d^2 k^2 \\
&= 0.
\end{aligned}
$$

Aida starts to solve this equation by collecting like powers of k and simplifying their coefficients. For k^2,

$$
\begin{aligned}
t_0^2 (b-a)^2 + d^2 (b+a)^2 - c^2 d^2 &= \\
t_0^2 (b-a)^2 + d^2 p t_0 &= \\
t_0^2 (b-a)^2 + t_0^2 t_1 t_2 &= \\
t_0^2 [(b-a)^2 + t_1 t_2] &= t_0^2 c^2.
\end{aligned}
$$

For k^1 (supplying the first step, which Aida omits),

$$
\begin{aligned}
-2d t_0 (b-a)^2 (b+a) &+ 2cd^2 D(b+a) - 4cd^2 v(b+a) \\
&- 2d^3 p(b+a) + 2c^2 D d^2 = \\
-2d t_0 (b-a)^2 (b+a) &+ 2cD d^2 p - 4cd^2 v(b+a) - 2d^3 p(b+a) \\
&= \\
-2d t_0 (b+a)[(b-a)^2 + d^2 p/t_0] &+ 2cD d^2 p - 4cd^2 v(b+a) = \\
-2d t_0 (b+a)[(b-a)^2 + t_1 t_2] &+ 2cd^2 Dp - 4cd^2 v(b+a) = \\
-2c^2 t_0 d(b+a) &+ 2cd^2 Dp - 4cd^2 v(b+a).
\end{aligned}
$$

For k^0,

$$-2cd^3Dp + d^2(b-a)^2(b+a)^2 - 4c^2d^2v(D-v) + 4cd^3vp$$
$$+ d^4p^2 =$$
$$d^2(b-a)^2(b+a)^2 - 4c^2d^2v(D-v) - 2cd^3p(D-2v) + d^4p^2$$
$$=$$
$$d^2(b-a)^2(b+a)^2 - c^4d^2 - 2cd^3p(D-2v) + d^2t_0t_1t_2p =$$
$$-c^2d^2t_1t_2 + c^2d^2t_0p - 2cd^3p(D-2v) =$$
$$-c^2d^2t_1t_2 + c^2d^2t_0p - \frac{c^3d^3p}{2v} + 2cd^3pv.$$

With the foregoing simplified coefficients, the key equation becomes

$$(c^2t_0^2)k^2 - [2c^2t_0d(b+a) - 2cd^2Dp + 4cd^2v(b+a)]k$$
$$- c^2d^2t_1t_2 + c^2d^2t_0p - \frac{c^3d^3p}{2v} + 2cd^3pv = 0.$$

Multiply this first by $1/c$:

$$ct_0^2k^2 - [2ct_0d(b+a) - 2d^2Dp + 4d^2v(b+a)]k - cd^2t_1t_2$$
$$+ cd^2t_0p - \frac{c^2d^3p}{2v} + 2d^3pv = 0.$$

Next, multiply by $2v$, noting that

$$-4d^2Dpv + 8d^2v^2(b+a)$$
$$= -d^2(4v^2 + c^2)(t_0 + 2c) + 8d^2v^2(t_0 + c)$$
$$= -c^2d^2p + 8d^2v^2(t_0 + c) - 4d^2v^2(t_0 + 2c)$$
$$= -c^2d^2p + 4d^2t_0v^2.$$

Aida, who does not mention the foregoing two steps explicitly, thus eliminates D and clears the denominator in the constant term to obtain

$$2ct_0^2 vk^2 - [4cdt_0 v(b+a) - c^2 d^2 p + 4d^2 t_0 v^2]k - 2cd^2 t_1 t_2 v$$
$$+ 2cd^2 t_0 pv - c^2 d^3 p + 4d^3 pv^2 = 0.$$

Finally, multiply through by $1/d^2$, taking advantage of the fact that doing so does not alter the quadratic form of the equation:

$$2ct_0^2 v \left(\frac{k}{d}\right)^2 - [4ct_0 v(b+a) - c^2 dp + 4dt_0 v^2]\left(\frac{k}{d}\right) - 2ct_1 t_2 v$$
$$+ 2ct_0 pv - c^2 dp + 4dpv^2 = 0.$$

The discriminant of the polynomial is

$$16c^2 t_0^2 t_1 t_2 v^2 - 16c^2 pt_0^3 v^2 + 8c^3 dpt_0^2 v - 32cdpt_0^2 v^3$$
$$+ 16c^2 t_0^2 v^2 (b+a)^2 - 8c^3 dpt_0 v(b+a)$$
$$+ 32cdt_0^2 v^3 (b+a) + c^4 d^2 p - 8c^2 d^2 pt_0 v^2$$
$$+ 16d^2 t_0^2 v^4.$$

Using the following four relations (again, not stated)

$$16c^2 t_0^2 t_1 t_2 v^2 - 8c^2 d^2 pt_0 v^2 = 8c^2 d^2 pt_0 v^2$$
$$16c^2 t_0^2 v^2 (b+a)^2 - 16c^2 pt_0^3 v^2 = 16c^4 t_0^2 v^2$$
$$32cdt_0^2 v^3 (b+a-p) = -32c^2 dt_0^2 v^3$$
$$8c^3 dpt_0^2 v(t_0 - b - a) = -8c^4 dpt_0 v$$

the discriminant becomes $8c^2 d^2 pt_0 v^2 + 16c^4 t_0^2 v^2 - 8c^4 dpt_0 v - 32c^2 dt_0^2 v^3 + c^4 d^2 p^2 + 16d^2 t_0^2 v^4$, which upon inspection is the square of $c^2 dp - 4c^2 t_0 v + 4dt_0 v^2$!

It is now a simple matter to apply the quadratic formula to obtain

$$\frac{k}{d} = \frac{4ct_0v(b+a) - c^2dp + 4dt_0v^2 + c^2dp - 4c^2t_0v + 4dt_0v^2}{4ct_0^2v}$$

$$= \frac{4ct_0^2v + 8dt_0v^2}{4ct_0^2v} = 1 + \frac{2dv}{ct_0}.$$

Hence $k = d + \frac{2d^2v}{ct_0}$, or, without d, $k = \frac{2t_1t_2v}{cp} + \sqrt{\frac{t_0t_1t_2}{p}}$. To facilitate computation of k when d is not given, Aida suggests letting $z = t_1t_2/p$ so that $k = \frac{2vz}{c} + \sqrt{t_0z}$.

Note that Aida's formula is equivalent to $\frac{k}{2} = r + \frac{1}{2} \cdot \frac{8r^2v}{2(s-c)c} = r + \frac{2r^2v}{[r^2s/(s-a)(s-b)]c} = r + \frac{2v(s-a)(s-b)}{sc}$, which is the form found in previous English-language sources.[17]

⚛

Aida's proof is a *tour de force* of algebra, but is short on geometrical motivation. A modern proof of the theorem in the form $\frac{k}{2} = r + \frac{2v(s-a)(s-b)}{sc}$ that gets straight to the nub of the matter follows from a wonderful theorem of Protasov (1992 in Russian; 1999 in English).

Protasov, who calls it the Segment theorem, establishes it in preparation for proving the celebrated theorem of Feuerbach, but

---

[17] This theorem is given as a problem in Fukagawa & Pedoe 1989. The proof in Fukagawa & Rigby 2002 omits many intermediate steps in the algebra; uses radii rather than diameters; gives expressions such as $s, s - c$, instead of p, t_0; and makes no use of altitude h. All these features suggest considerable editing of some unidentified original Japanese text. Although Fukagawa and Rigby ascribe the theorem to Ajima Naonobu, they do not mention the name of the source text; so far, I have found this problem in Japanese only in works by Fujita Yoshitoki and Aida. For another approach to the proof, see Unger 2010.

the Segment theorem by itself turns out to be exactly what is needed to prove the Japanese theorem. As Protasov's proof of the Segment theorem is quite lengthy and readily accessible, I merely explain its substance here and show how it leads directly to the Japanese theorem.

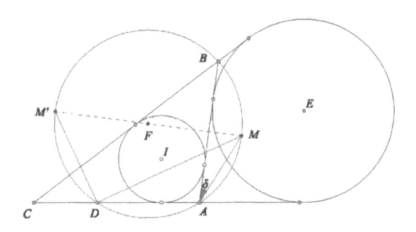

Take an arbitrary triangle ABC with incircle (I) and excircle (E) on side AB. Pick a point D anywhere on line AC—between A and C, as in the figure, or beyond either endpoint of segment AC—and construct the circle (F) determined by points A, B, D. The line through F perpendicular to AB meets (F) in the midpoints M, M' of the two arcs $\overset{\frown}{AB}$. Since $\overset{\frown}{BM} = \overset{\frown}{MA}$, $\angle MAB = \angle ADM = \angle BFA/4$. We shall call this angle δ. Since $DM \perp DM'$, $\angle CDM'$ is its complement.

Suppose now that we construct $HI \perp DM$ with H on AC and $GH \perp AC$ with G on CE. Protasov proves *en passant* that this makes $\angle GHI = \delta$. Moreover, $(G)H$ is a pseudo-incircle of the triangle ABC and circle (F). In fact, Protasov proves more generally that the perpendiculars to DM and DM' through I and E meet CA in the four points touched by the circles that are tangent to CA, CB, and (F).[18] This is his Segment theorem. Two circles touch (F) externally and two (of which (G) is one) touch it internally.

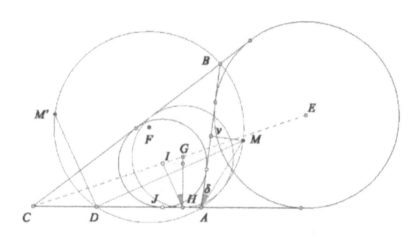

The connection with the Japanese theorem is this: Say that (I) touches AC at J. In the customary way, we designate $\angle ACE$ as $\gamma/2$, inradius IJ as r, and semiperimeter $\frac{a+b+c}{2}$ as s. Clearly $\tan\frac{\gamma}{2} = \frac{r}{s-c}$ and $\tan\delta = \frac{v}{c/2}$. (This much is stated in Fukagawa & Rigby 2002.) From $rs = \sqrt{s(s-a)(s-b)(s-c)}$, we have $\frac{r}{s-c} = \frac{(s-a)(s-b)}{rs}$. Finally, $GH - IJ = HJ\tan\frac{\gamma}{2}$.

---

[18] As Protasov explains, Feuerbach's theorem is the special case of the Segment theorem that arises when $\angle BAC = \pi - 2\delta$ and (F) is the Nine Point Circle of ABC.

Now $HJ = r \tan \delta$ and $IJ = r$, so $GH - r = r \tan\frac{\gamma}{2}\tan \delta = r \cdot$ $\frac{2v}{c} \cdot \frac{(s-a)(s-b)}{rs} = \frac{2v(s-a)(s-b)}{sc}$. This is equivalent to the Japanese theorem because $GH = \frac{k}{2}$ and $\angle GHI = \delta$ for any position of D on AC. One cannot help suspecting that Aida and other Japanese (or at least the one who discovered the theorem) must have understood, at least intuitively, that $\angle GHI = \angle BAM$ for all choices of D, and that the lengthy algebraic proof came later.

Problem 7a

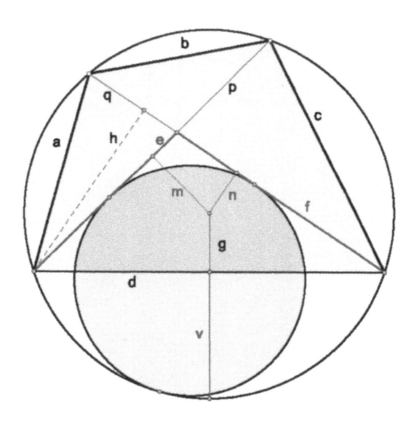

A quadrilateral with sides a, b, c, d and diagonals p, q is inscribed in a circle of diameter D, **the center of which lies within the quadrilateral**. What is the diameter k of the pseudo-incircle of the triangle consisting of the portions e, f of the diagonals marked in the figure and circumcircle of the quadrilateral? ANSWER: if $a = 39, b = 25, c = 52, d = 60$, then $D = 65, p = 56, q = 63$, and $k = 40$.

METHOD: Aida refers the reader to fascicle 12 of his book *Sanpō kantsū jutsu*, in each fascicle of which he analyzes a class of problems in increasing order of complexity. In fascicle 12, the first problem asks for the diameter of the circumcircle of a triangle given its sides; the second, for the diameter of the circumcircle of a cyclic quadrilateral; third and fourth, for the lengths of the diagonals of a cyclic quadrilateral; the fifth, for the lengths of the diagonals of a cyclic pentagon; and so on. The first four problems establish a theorem succinctly proven in modern form by Alsina & Nelsen 2007:

If a, b, c, d denote the lengths of the sides; p, q the lengths of the diagonals; R the circumradius; and Q the area of a cyclic quadrilateral, then

$$p = \sqrt{\frac{(ac + bd)(ad + bc)}{ab + cd}},$$

$$q = \sqrt{\frac{(ac + bd)(ab + cd)}{ad + bc}},$$

and

$$Q = \frac{1}{4R}\sqrt{(ab + cd)(ac + bd)(ad + bc)}.$$

Letting $ab + cd = s$, $ad + bc = t$, $ac + bd = u$, and $D = 2R$, this is equivalent to $p = \frac{\sqrt{tu}}{\sqrt{s}}$, $q = \frac{\sqrt{su}}{\sqrt{t}}$, and $2DQ = \sqrt{stu}$. Aida writes p and q just this way, but chooses to write $D^2 = \frac{4s^2p^2}{4s^2 - (c^2 - a^2 + d^2 - b^2)^2}$, in which the denominator can be factored into

$$(-a + b + c + d)(a - b + c + d)(a + b - c + d)(a + b + c - d)$$
$$= 16Q^2$$

(by Brahmagupta's formula), and the numerator equals $4stu$. Thus $D^2 = \frac{stu}{4Q^2}$.

An advantage of expressing 16 times the area of the quadrilateral as the difference of two squares becomes apparent in the next step, in which Aida successively factors the left side of the identity

$$4s^2p^2 - 4s^2D^2 + (c^2 - a^2 + d^2 - b^2)^2D^2 = 0$$

as

$$-4s^2(D^2 - p^2) + (c^2 - a^2 + d^2 - b^2)^2D^2 = 0$$
$$-16s^2m^2 + (c^2 - a^2 + d^2 - b^2)^2D^2 = 0$$
$$-4sm + (c^2 - a^2 + d^2 - b^2)D = 0$$

and thus obtains $m = \frac{D(c^2 - a^2 + d^2 - b^2)}{4s}$, which is much simpler than the more direct $m = \frac{1}{2}\sqrt{D^2 - p^2}$. Correspondingly, one can show that $n = \frac{D(c^2 - a^2 + d^2 - b^2)}{4t}$, though Aida does not write out a proof.

He next states that $e = \frac{adp}{t}$ and $f = \frac{cdq}{s}$, referring the reader to a supplement to *Sanpō kantsū jutsu* for justification. I have been unable to locate this work, but no doubt Aida was thinking of similar triangles. Because $\frac{e}{f} = \frac{a}{c}$ and $\frac{e}{q-f} = \frac{d}{b}$, we have $\frac{e^2}{f(q-f)} = \frac{ad}{bc}$. Because $f(q - f) = e(p - e)$ (Crossed Chords theorem), this means that $\frac{e^2}{f(q-f)} = \frac{e}{p-e} = \frac{ad}{bc}$. Likewise, $\frac{f}{q-f} = \frac{cd}{ab}$. Solving these yields Aida's expression for e and f.

Aida does not explain $h = \frac{ad}{D}$, but it follows, in modern terms, from the Law of Sines, according to which $\frac{a}{\sin \alpha} = \frac{d}{\sin \delta} = D$, where α and δ are the angles opposite sides a and d, respectively, formed by them and diagonal q. Multiplying, $\frac{ad}{\sin \alpha \sin \delta} = D^2$ or $\frac{ad}{h^2/ad} = D^2$.

Because, adding triangle areas, $\frac{1}{2}fh = \frac{1}{2}em + \frac{1}{2}fn + \frac{1}{2}dg$, we know $g = \frac{fh}{d} - \frac{em}{d} - \frac{fn}{d}$, and consequently the sagitta $v = \frac{D}{2} - g$. Finally, denoting the diameter of the incircle of the critical rectilinear triangle as w, we have $\frac{w}{2} \cdot \frac{d+e+f}{2} = \frac{1}{2}fh$, so $w = \frac{2fh}{d+e+f} = \frac{2adf}{D(d+e+f)}$. Since we can express d, e, f, w, and v in terms of a, b, c, d, we can apply the formula discussed in the remarks on problem pair 6, which now becomes $k = w + \frac{2w^2v}{d(e+f-d)}$, and is the solution Aida gives for 7a.

Problem 7b

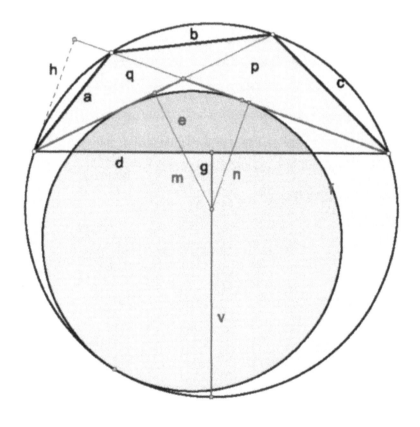

A quadrilateral with sides a, b, c, d and diagonals p, q is inscribed in a circle of diameter D, **the center of which lies outside the quadrilateral.** What is the diameter k of the pseudo-incircle of the triangle consisting of the portions e, f of the diagonals marked in the figure and the circumcircle of the quadrilateral? Answer: if $a = 18.2 \, b = 16, c = 25, d = 52$, then $k = 57\frac{13}{17}$.

METHOD: The solution is the same as that of 7a, but $v = \frac{D}{2} + g$ instead of $v = \frac{D}{2} - g$, and $g = -\frac{fh}{d} + \frac{em}{d} + \frac{fn}{d}$.

REMARKS ON PROBLEM PAIR 7

In *Sanpō kantsū jutsu*, the exposition is more complicated than the one just given. Aida first shows how to find D given b, d, p, q. He shows how to determine a, c, k from the same givens. Next he shows how to find m, n, g given a, b, c, d. Finally he addresses Problem 7a, and goes on to consider the three other pseudo-incircles that could be added to the figure. By contrast, in *Kajiten sandaishū*, Aida explains 7a directly.

In giving the solutions to 7a and 7b, Aida suggests introducing additional placeholders to simplify the calculation. In giving the introductory answer to 7b, he mistakenly writes $p = 32$ for $p = 33$, and in a summary table of values after the solution, he has $a = 18.2, b = 18.2, c = 16, d = 52$, an obvious miscopying of $a = 18.2 \, b = 16, c = 25, d = 52$. Nonetheless, in the table, he correctly gives

$$q = 39, e = 23\frac{10}{51} \approx 23.1961, f = 31\frac{44}{51} \approx 31.8627,$$
$$m = 28, n = 26, g = 19.5, h = 14.56,$$

and $v = 52$. The correct value for k is 57.7778, but Aida writes $57\frac{13}{17} \approx 57.7647$.

Problem 8a

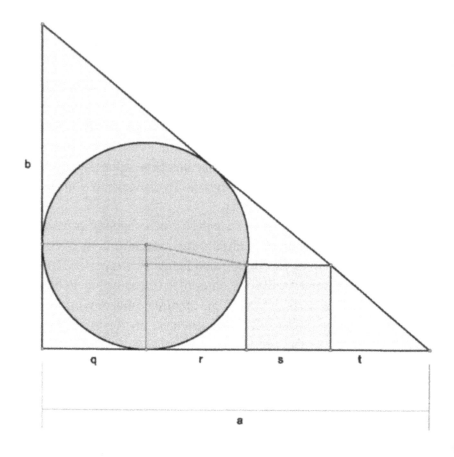

A square is placed in a right triangle with one side on leg a, one vertex on the hypotenuse, and the fourth side touching the incircle. **The top of the square lies below the center of the circle.** If $a = 4$ and $b = 3$, what is the side s of the square?

METHOD: Since a and b are legs of a right triangle, we know that $d/2 = a + b - \sqrt{a^2 + b^2}$. Furthermore, in the figure,

$$q = \frac{d}{2}, \qquad r = \sqrt{\left(\frac{d}{2}\right)^2 - \left(\frac{d}{2} - s\right)^2} = \sqrt{s(d - s)}, \qquad t = s\frac{a}{b}.$$

Therefore, since $a - q - s - t = r$, we have

$$a - \frac{d}{2} - s - \frac{as}{b} = \sqrt{s(d - s)}.$$

Multiplying through by b and squaring both sides yields

$$\left[b\left(a - \frac{d}{2}\right) - s(b + a)\right]^2 = b^2 s(d - s).$$

That is,

$$b^2\left(a - \frac{d}{2}\right)^2 - 2b(a + b)\left(a - \frac{d}{2}\right)s + (a + b)^2 s^2 = b^2 s(d - s),$$

or, grouping together powers of s as Aida prefers,

$$b^2\left(a - \frac{d}{2}\right)^2 - \left[b^2 d + 2b(a + b)\left(a - \frac{d}{2}\right)\right]s + [b^2 + (a + b)^2]s^2$$
$$= 0. \ \square$$

Problem 8b

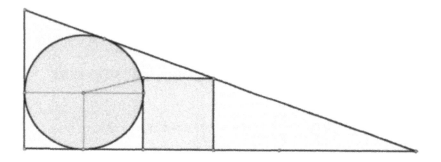

A square is placed in a right triangle with one side on leg a, one vertex on the hypotenuse, and the fourth side touching the incircle. **The top of the square lies above the center of the circle.** If $a = 12$ and $b = 5$, what is the side s of the square? ANSWER: $s = 2\frac{6}{17}$. The diameter of the incircle is 4.

METHOD: The derivation is short but a bit cryptic. Aida defines c as $\frac{ab}{a+b}$ and asserts that $\frac{a}{a-d} = \frac{c}{s}$, which implies $s = \frac{c(a-d)}{a}$. Hence $\frac{b(a-d)}{a+b} = s$. $\square$

REMARKS ON PROBLEM PAIR 8

I have followed Aida in not labeling the segments in figure 8b. He implicitly carries over the designations of lengths from Problem 8a.

In 8a, the endpoint of the fourth side of the square not on leg a touches the incircle, but this is of no consequence for Aida. No numerical answer is provided, and no summary recipe for the calculation is given. All we have is the quadratic for finding s.

In 8b, the c does not appear to have a geometric correlate in the figure, but the result is correct. Using the segment t defined in 8a, $a = d + s + t = d + s + \frac{sa}{b}$ by similar triangles, from which it follows at once that $\frac{b(a-d)}{a+b} = s$.

In the copy of *Kajiten sandaishū* in the Sakuma Collection at Yamagata University, some pages (which are numbered) have been resewn into the fascicle (presumably when the binding was being repaired) in the wrong order. Our Problems 8a, 8b, 9a, and 9b appear in the order 8a, 9a, 8b, 9b.

Problem 9a

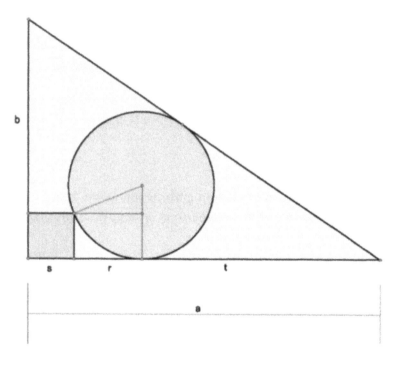

A right triangle with sides a, b, c is filled with a circle and a square as shown. **The top of the square lies below the center of the circle.** If $a = 84$ and $b = 63$, and the diameter of the circle is $e = 40$, what is the side of the square? ANSWER: $s = 8$.

METHOD: Denoting the diameter $a + b - \sqrt{a^2 + b^2}$ of the incircle (not shown in Aida's figure) as d, we have

$$t = \frac{e\left(a - \frac{d}{2}\right)}{d}.$$

Since $r = a - t - s$ and $r^2 = s(e - s)$ (by the Crossed Chord theorem), we have $(a - t - s)^2 = s(e - s)$. This expands to $(a - t)^2 - 2s(a - t) + s^2 - es + s^2 = 0$, or, as a quadratic in s,

$$(a - t)^2 - [2(a - t) + e]s + 2s^2 = 0.$$

The discriminant simplifies to $\Delta = -4(a - t)^2 + 4e(a - t) + e^2$, and, as we seek $s < e/2$, we choose the root with $-\sqrt{\Delta}$ in the quadratic formula. Aida gives the result as $0 = -2(a - t) - e + \sqrt{\Delta} + 4s$. □

There is a short continuation of the text, which I omit: see the Remarks on this problem pair.

Problem 9b

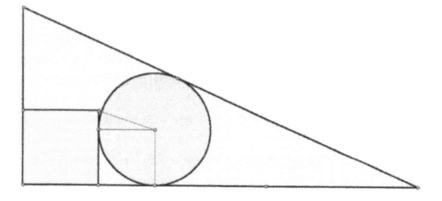

A right triangle with sides a, b, c is filled with a circle and a square as shown. **The top of the square lies above the center of the circle.** If $a = 4$ and $b = 3$, and the diameter of the circle is $e = 1.4$, what is the side of the square? ANSWER: $s = 1.2$.

METHOD: Letting d be the diameter of the incircle (since this is a 3-4-5 right triangle, $d = 2$), we have

$$\frac{a}{a-s} = \frac{d}{e},$$

so $d(a - s) = ae$ or $0 = ds - ad + ae$. I.e.,

$$s = a - \frac{ae}{d}. \quad \square$$

REMARKS ON PROBLEM PAIR 9

Aida does not label the segments in figure 9b

The equation for t in 9a assumes an auxiliary line (not shown) touching the circle parallel to side b so that the given circle is the incircle of the resulting triangle. Then, by similar triangles, we get $a - d/2 : d :: t : e$, which leads to the formula for t.

On the page after the solution of 9a, we find two equations that use the quantities $f = \frac{ae}{d}$, $g = a + \frac{e}{2} - f$, and $h = \sqrt{f^2 - 2g^2}$. These are $g^2 - 2fs + 2s^2 = 0$ and $h - g + 2s = 0$. The first is probably a mistake for $g^2 + 2(f - a - e)s + 2s^2 = 0$, which is equivalent to the solution equation with t expressed in terms of a, d, e. Likewise, the second equation seems to be an attempt to write $0 = -2(a - t) - e + \sqrt{\Delta} + 4s$ without t. Since Aida obviously "cooked" the data to yield integer roots (s is 8 or 36), these equations were probably just an afterthought. Could they perhaps have been added by a student while copying the manuscript?

Problem 10a

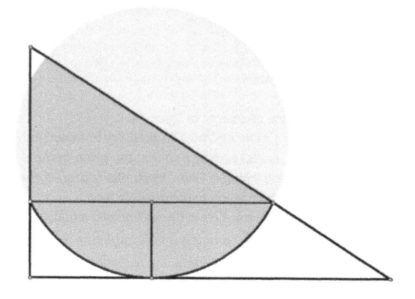

A circle touches one and only one leg of a right triangle. The other leg and hypotenuse intercept a chord of length k. The sagitta on this chord has length s less than the radius of the circle. If $a = 4$, $b = 3$, and $s = 0.6$, what is k? ANSWER: $k = 3.2$.

METHOD: By similar triangles, $k = \frac{a(b-s)}{b}$. $\square$

Problem 10b

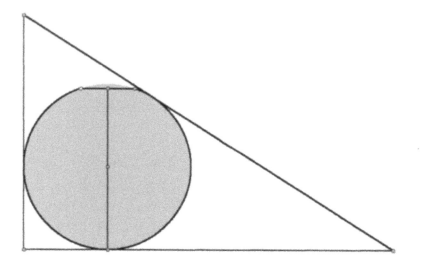

A circle touches all three sides of a right triangle. There is a chord of length k parallel to one leg; the length of the sagitta on this chord is s. If $a = 4, b = 3$, and $s = 1.8$, what is k? ANSWER: $k = 1.2$.

METHOD: The circle is obviously the incircle. By the Crossed Chords theorem, $k^2 = 4s(d - s)$, where d is the diameter of the incircle. Because we have a right triangle, we can substitute $a + b - \sqrt{a^2 + b^2}$ for d and find k. $\square$

Problem 10c

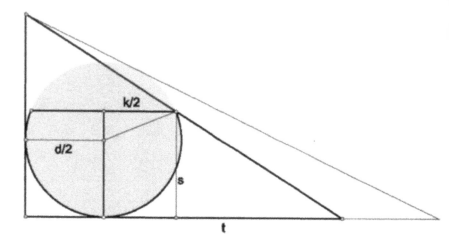

A circle larger than the incircle touches both legs of a right triangle. A chord of length k parallel to one leg passes through one of the points in which the hypotenuse meets the circle. The sagitta has length s greater than the circle's radius $d/2$. If $a = 4$, $b = 3$, and $d = 2.5$, what is the length of the sagitta? ANSWER: $s \approx 1.5057 \dots$.

METHOD: By the Crossed Chords theorem, $(d - s)s = \left(\frac{k}{2}\right)^2$. On the one hand, $t = a - \frac{d}{2} - \frac{k}{2}$. On the other, $\frac{b}{s} = \frac{a}{t}$ or $as - bt = 0$. Therefore $as - b\left(a - \frac{d}{2}\right) + \frac{bk}{2} = 0$.

Rearranging and squaring,

$$a^2 s^2 - 2abs\left(a - \frac{d}{2}\right) + b^2\left(a - \frac{d}{2}\right)^2 = \frac{b^2 k^2}{4}$$

$$a^2 s^2 - 2abs\left(a - \frac{d}{2}\right) + b^2\left(a - \frac{d}{2}\right)^2 = b^2(d - s)s$$

$$a^2 s^2 - 2abs\left(a - \frac{d}{2}\right) + b^2\left(a - \frac{d}{2}\right)^2 = b^2 sd - b^2 s^2$$

or, as a quadratic in s,

$$b^2\left(a - \frac{d}{2}\right)^2 - \left[2ab\left(a - \frac{d}{2}\right) + b^2 d\right]s + (a^2 + b^2)s^2 = 0. \quad \square$$

REMARKS ON PROBLEM SET 10

This is the only trio of problems in *Kajiten sandaishū*.

Aida's auxiliary figure for Problem 10c includes extra lines showing the triangle of which the given circle is the incircle. For $a = 4$ and $b = 3$, if $d = 2.5$, then $s = 1.1294 \dots$, which is too small for $s > \frac{d}{2}$, or 2.4106 ... , which is not the purported answer. If $s = 1.5057$ ... , then $d = 11.8951$ ... , which is too large, or 2.0973 ... . Evidently something has been miswritten or miscopied.

Suspicion first falls on the peculiar expression for s, 一寸五分 令五七 有奇, which I have interpreted above as 1.5057 ... , taking 令 as clipped form of *rei* 'zero' 零 and treating the whole string as a decimal fraction. The only alternative I can think of, judging from other passages I have encountered in Aida's works, is $s = 1 + \frac{0.57}{5} = 1\frac{57}{500} \approx 1.14$, which yields $d = 2.5233$ ... and a second, larger root, but this violates the condition $s > \frac{d}{2}$. So, accepting $s = 1.5057$ ... , it seems that Aida (or someone else) wrote $d = 2.5$ 一 寸五分 for $d = 2.1$ 一寸一分, for which $s = 1.5015$ ... .

As in 9a, a discriminant and linear equation for s are given immediately after the solution quadratic in 10c. Neither seems to be correct given the solution equation, strengthening the impession that we are dealing with some kind of scribal error.

CRSƆ

Though this is the last full entry in *Kajiten sandaishū*, the last few pages of the fascicle contain a short envoi. It consists of eight figures, arranged in pairs involving scalene triangles, the first of which is the following.

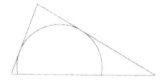 

The center of the circular arc on the left lies above the base of the triangle, whereas the center of the arc on the right lies below it. The remaining pairs show other differences:

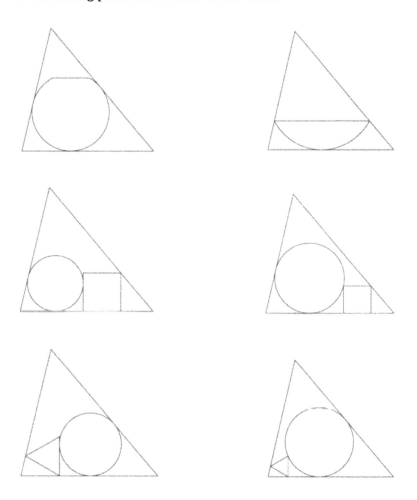

Aida states that, in each pair, the constants and method of solution differ. He also observes that more pairs of this sort can be produced *ad libitum*, using isosceles triangles, pentagons, hexagons, and so on.

Metrical solutions of these problems would obviously require more effort than they would if the triangles were right, as they are in sets 8, 9, and 10, instead of scalene. Did Aida mean to propose such problems as challenges to his students? Did he entertain ideas about a general method for tackling such problems? We cannot be sure, but, once again, they suggest that *Kajiten sandaishū* was a book to be used in teaching.

Appendix A: Aida as a Student

Every teacher once had a teacher, and in Aida's case, the most influential was probably his near-contemporary Honda Toshiaki. Honda (1744–1821) was an extraordinarily bold and independent intellectual: a *wasanka*, astronomer, ship's captain, political economist, and scholar of *rangaku* or Dutch studies (Keene 1969; Stephan 1983; Horiuchi 2002). He made it his business to become as well-informed about the nations of Europe and their technologies as he could during an age of officially enforced national isolation, and to advocate (unsuccessfully) for policy changes based on his knowledge.

In 1993, Kobayashi Tatsuhiko published an article about a one-fascicle work by Aida called *Oranda sanpō* Dutch Calculation. In the article, Kobayashi explains in detail how Honda's explication of the Dutch use of plane trigonometry in navigation informed Aida's work, and discusses the difference between Honda's and Aida's view of mathematics.[19] In addition, because Kobayashi refers to numerous manuscripts by Honda, Aida, and other *wasanka* as well as later secondary studies by historians of Japanese mathematics, his article gives the reader a good sense of the range and depth of academic scholarship on *wasan*. By comparison, my conjecture that Aida's *Kajiten sandaishū* was meant as teaching material rests on a much more modest foundation of internal features of the text.

Those who can read Japanese can read Kobayashi's article online for themselves (see the References for a link), and translating

---

[19] Neither Honda nor Aida mention spherical trigonometry (on the history of which see Van Brummelen 2013: 73 ff.) in the texts discussed by Kobayashi, but, in personal communication, Kobayashi, who has subsequently researched the influence of Chinese translations of Western texts on calendrical calculations, avers that, by the mid 18th century, it was known in Japan.

it with sufficient annotation to make it helpful to those who do not read Japanese or are only slightly acquainted with *wasan* would make this book too large. I therefore limit myself in this appendix to comments that give the reader a feeling for the variety of scholarly questions that arise when studying works of *wasan*. My goal is to help the reader see that they require attention to aspects of premodern Japanese history that go beyond the mathematics of the problems recorded on the colorful *sangaku* that were displayed in Edo period shrines and temples, including contemporary networks of communication.

෬෫

According to Smith and Mikami (1914: 263), *Oranda sanpō* dates from about 1790, but Kobayashi says only that the latest possible date is 1806. Furthermore, the text exists in several manuscript versions. I consulted the digitized version in the Sakuma Collection of Yamagata University Library. Kobayashi examined manuscripts in several different institutional and personal collections in Japan. Why are the date of composition and original wording of the text important?

Take the date issue first. Japanese measured large distances in *ri* 里, a unit known to have been about 660 meters when it was introduced from China,[20] but which came to be fixed at 3,927 meters in the Edo period. Kobayashi points out that an urgent question for Japanese mathematicians at the dawn of the 19th century

---

[20] Evidence for this comes from old names, the best known of which is Kujūkuri '99 *ri*' 九十九里 for a particular beach on the Pacific side of modern Chiba prefecture. It is about 65 km long, and the name is said to have been conferred on the place after the first Kamakura *shōgun* Minamoto Yoritomo ordered that it be measured. A fishing scene there is pictured in the 19th multicolor woodblock print in the series *Famous Places in the Sixty-Odd Provinces* by Utagawa Hiroshige (1797–1858).

was the number of *ri* in one degree of latitude. He states that, in a work of 1804, Honda gave the number as 30, but as 32 in another work; Aida, in two places, gave the figure 25. The discrepancy, noteworthy in itself, is important because the actual distance is about 111 km, which is within 1 km of 28.2 *ri*, the figure arrived at by Inō Tadataka in 1801 and corroborated by his teacher Takahashi Yoshitoki in the following year. The issue here, of course, is not one of metrical accuracy but rather of philology: if both Honda and Aida came up with their estimates after Inō had settled on his, as Kobayashi thinks, then their failure to mention it may tell us something about the time it took for information to spread within *wasan* circles of the early 1800s. On the other hand, if Aida wrote *Oranda sanpō* around 1790, as Smith and Mikami believed, then there must be some other reason for his lower estimate.

CRSO

Consider next the matter of manuscript variants. Although woodblock book publishing was thriving at the end of the 18th century in Edo and Kyōto (Kornicki 1998), the usual method for dissemination of *wasan* material was the hand-copying of manuscripts (see p. 7 above). The manuscript of *Oranda sanpō* I have studied has a curious feature apparently absent in the version Kobayashi relied on: the last six pages give two different versions of the same problem (three pages each). They differ in both numerical data and narrative wording. For three reasons, we would really like to know which version of the problem, the finale of *Oranda sanpō*, was the original. First, as Kobayashi shows, both versions seem to be derived from a problem given and solved by Honda using Dutch methods. Second, Aida's version of the problem in the form quoted by Kobayashi involves absurdly large distances. Finally, the version not quoted by Kobayashi raises new questions about the composition of *Oranda sanpō*.

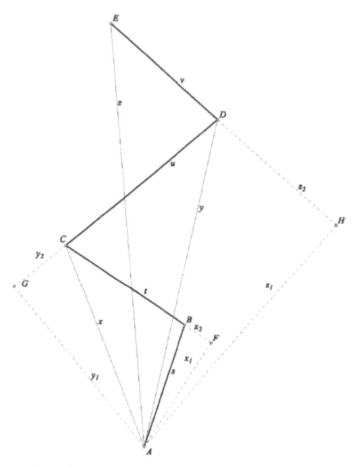

Honda's problem involves calculations of the kind often called dead reckoning.[21] Assuming a flat ocean, suppose a ship sails the path $ABCDE$ in the figure above and that we know the distances s, t, u, v and headings or AZIMUTHS (degrees clockwise from North)

---

21 *Dead* here is a product of false etymology. The phrase *deduced reckoning* was shortened to *ded reckoning*, and misspelling followed.

of each segment (AB, BC, CD, DE) of the journey. Using this information, one can determine, with the help of trigonometry, the heading of the straight-line course AE and its length. (I give Kobayashi's explanation with some gaps filled in.)

From the headings, we can compute $\angle ABC, \angle BCD, \angle CDE$. Constructing $AF \perp BC, AG \perp CD, AH \perp DE$ as shown, our plan is to work our way up the resulting chain of right triangles starting with AFB, and find x, y, z in turn. Looking at triangle AFB, we see that $x_1 = s \sin \angle ABF$ and $x_2 = s \cos \angle ABF$. Since $\angle ABF$ is the supplement of $\angle ABC$, we can easily compute x_1 and $x_2 + t$; the Pythagorean theorem then gives us $x = AC$.

Moving onto y, we see that $\angle BCG$ is the supplement of $\angle BCD$, which we know. Applying the Law of Cosines to triangle ABC (we know its sides s, t, x), we compute $\angle ACB$. Now $\angle BCG - \angle ACB = \angle ACG$ and we have $y_2 = x \cos \angle ACG$. Two more applications of the Pythagorean theorem give us to y, viz. $x^2 - y_2^2 = y_1^2$ and $y_1^2 + (y_2 + u)^2 = y^2$.

We can now find the distance z and its azimuth using the same technique we used to find x. In principle, the number of segments can be increased, and Honda rightly called this *kukkyoku tokai* 'zigzag navigation' 屈曲渡海.

Aida, unlike Honda, does not explain the trigonometry involved. A translation of the version given at the end of the Yamagata manuscript of *Uranda sanpō*, which is very close to the version quoted by Kobayashi, runs as follows:

The Technique of Measuring Distances at Sea

Suppose that, from a port in a country in the East, there is another country off in the West, which no one can see, but to which, according to tales of old, people go when they die, traveling there in the Boat of Salvation.

Now, the sea lies on the surface [lit. skin] of the earth, so it forms a sphere. Hence it is impossible for anyone to see a place tens of thousands of *ri* in the distance; indeed, they say a ship's mast can be seen [only] as far as seven *ri* from shore. Therefore, it is impossible to figure out the distance to a place tens of thousands of *ri* away. To do so, one uses the technique of dead reckoning.

Because it is hard to maintain a straight course and there are many chances to be led astray by demons, it is necessary to use calculations if one wishes to return to one's starting point after having traveled to a distant land. Suppose you take a heading of 264° [clockwise from North][22] for 60,000 *ri*. You can't see anything in any direction at that spot. You now take heading 180° [due South] for 37,500 *ri* with the same result. Next, you take heading of 342° for 60,000 *ri*; again, can see nothing, so you go on heading 183° for 45,000 *ri*. Still nothing can be seen. Going 291° for 15,000 *ri*, you arrive at last at land. What is the direction from the port of arrival back to the port of departure and its distance along a straight line? The answer is 99,000 *ri* on a heading of 75°. One uses short distances to calculate large ones.

Kobayashi notes the obvious references to magic and the Pure Land of popular Buddhism in this version of the text without comment, but he does not hesitate to characterize Aida's grasp of geography, shown by the enormous distances he gives, as impoverished. Since, for Aida, the circumference of the earth

---

[22] Although systems based on four or eight directions existed, when Japanese of the Edo period needed to specify directions precisely, they most often used twelve sectors of 30° named after the twelve signs of the Chinese zodiac. Each sector was divided into tenths, each measuring 3°. Thus *shi gobun* '1st sign, 5/10' 子五分 was North (0°), the whole sector running clockwise from 345° to 15°. The direction with azimuth 21° was *chū nibun* '2nd sign 2/10' 丑二分, since the next sector ran from 15° to 45°; and so on.

would have been 9,000 *ri* assuming 25 *ri* for each degree of latitude (the figure mentioned above), the ship in the problem would
have to circumnavigate the globe eleven times to get home!

But consider the version of the problem in the Yamagata manuscript that Kobayashi did not quote. I have used boldface to highlight important differences from the second version. It has no title.

Suppose that, from a port in a country in the East, there
is **an island to the Northeast**, which no one can see, but
**over which clouds gather when, from time to time, a
typhoon occurs, giving a hint that it is there**.

Now, the sea lies on the surface [lit. skin] of the earth,
so it forms a sphere. Hence it is impossible for anyone to
see a place **ten *ri*** in the distance; indeed, they say a ship's
mast can be seen [only] as far as seven *ri* from shore.
Therefore, it is impossible to figure out the distance to a
place **hundreds** of *ri* away. To do so, one uses the technique of dead reckoning.

**Without such techniques of calculation, one will
surely not be able to find the way back. Hence one
wishes to know where the island is in terms of those
calculations.** Suppose you take a heading of **24°** [clock-
wise from North] for **400** *ri*. You can't see anything in
any direction at that spot. You now take heading **290°** for
250 *ri* with the same result. Next, you take heading of
102° for **400** *ri*; again, can see nothing, so you go on
heading **303°** for **300** *ri*. **Now you think you see something off to the East, and** going **51°** for **100** *ri*, you arrive at last at **an island**. What is the direction from the
port of **departure** to **this island** and its distance along a
straight line? The answer is **660** *ri* on a heading of **15°**. [23]

---

[23] There are also slight differences between the Yamagata manuscript
and the version transcribed in Kobayashi's article. For instance, he
writes *manjutsu* 'all manner of techniques' 万術 for a word I found as 亍
術, a variant of 算術 'calculation techniques'. But I pass over such details
as they would take us too far afield.

This version of the problem seems far more reasonable. Why was it not in the manuscript Kobayashi studied? Who wrote which version of the problem in the Yamagata fascicle, and when? Depending on the answers to such philological questions, we might conclude that Aida's geographical understanding was not quite so poor after all. They might even shed some light on another question that Kobayashi raises: how did Aida estimate the limits of vision along a straight line and determine that ships' masts vanish below the horizon at a distance of seven *ri* from shore? In his 1993 article, Kobayashi says he had yet to find a passage in one of Honda's works that could be the source, but perhaps Aida did the calculation on his own (see again note 20 above). Honda championed the Copernican system, which, unlike the traditional Chinese idea of the world as a disk "under heaven,"[24] entails the idea of the earth as a rotating ball. As Kobayashi rightly says, we should not be surprised to find Aida saying that the earth is round.[25]

CRSO

Oranda sanpō contains another additional non-mathematical sign that Aida relied on information he had picked up from Honda. In his preface, Aida refers to a Dutch book he calls *Seiheirukonsuto* セイヘイルコンスト, later writing this as *Zeiheirukonsuto* ゼイヘイルコンスト. He explains that this means 'the art of crossing the

---

[24] Indeed, Ch *tiānxià* (J *tenka*) 'under heaven' 天下 meant the universe, of which China was literally *zhōnghuá* 'the flower at the center' 中華. In the course of time, the word written 天下 was put to other political uses in Japan, Korea, and Vietnam.

[25] Actually, Aida seems to be saying more than that the earth is round. Did his use of 'skin' perhaps imply that the irregularities of mountains and valleys we see on land continue out into the ocean, the surface of which is smoother because water seeks its own level?

seas', which suggests something like modern Dutch *Zeevaartkunst* 'art of navigation' or, in older spelling, *Zeevaertkonst.* Here Kobayashi's familiarity with other primary source materials comes into play. He explains that the *wasanka* Satō Setsuzan (1814–1859), who, like Honda, hailed from Echigo (modern Niigata),[26] mentions a Dutch work he recorded as *Talkunstboek of Cÿhelkunsboek* 'book on the art of numbers or book on the art of ciphering'. He transcribed the last word, more properly spelt *Cijpherkunstboek*, as *seeherukonsutobukku* セーヘルコンストブック, which shows a good fit with Aida's *seiheirukonsuto.* Kobayashi therefore suggests the possibility that Aida's source may have been a book on algebra rather than navigation. This is brilliant detective work, and if Kobayashi is right, we are forced to conclude that Aida's knowledge of Dutch was strictly vicarious. Perhaps he was confusing two books he had seen or things he had been told about at different times.

Returning to mathematics, Kobayashi concludes that, though Aida did not hesitate to rely on this teacher's findings, he was not as receptive to the unfamiliar ideas of the Hollanders. Whereas Honda saw trigonometry as something fellow Japanese would do well to learn and adopt,[27] Aida's instinct was to recast Dutch methods in terms familiar to *wasanka.* Kobayashi, whose primary goal is to elucidate the relationship between Honda and Aida, illustrates this with a single quotation from *Oranda sanpō* in which Aida complains that Dutch methods involve approximations and

---

[26] Echigo bordered on the province of Dewa, the southern part of which in modern times has become Yamagata prefecture. Aida was born there in the town of Mogami 最上. The same *kanji* binom can also be read *saijō* 'highest', and Aida adopted this graphic pun as the name of his school of *wasan.*

[27] Honda had many other radical suggestions, including the replacement of Chinese characters with the roman alphabet, the expansion of Japan as far north as Sakhalin (J Karafuto), and the establishment of a capital city there on the same latitude as London!

therefore are not as good as the exact solutions to various problems. I have no quarrel with this overall conclusion, but I think there are additional features of *Oranda sanpō* that Kobayashi does not mention but are worthy of comment.

For instance, Aida says that the Dutch divided a circle into 360 degrees or 21,600 minutes, yet seems wedded to the idea of the measure of an angle as the ratio of the length of the arc it subtends in some actual circle to the diameter of that circle. Guessing that 360 degrees was an ancient approximation to the 365¼ days of the (Julian) year, Aida gives a table showing the diameter d_s of the annual circuit of the sun (taking the earth as stationary) for 365¼, 365, 360, and 120 degrees, and the diameters d_e of the earth for a circumference c of 90,000 or 1,500 *ri*. Why does he not simply say that the ratio of circumference to diameter is always π? That is the fundamental reason, after all, that trigonometric tables are useful. By emphasizing the arbitrary size of degrees and *ri*, Aida seems to be justifying his use of actual arc lengths in his later discussion of Dutch methods.

The first two problems Aida discusses highlight this aspect of his thinking. They are also interesting because they are variants of essentially the same problem. (If *Kajiten sandaishū* is indeed teaching material, it seems that contrasting closely related problems was one of Aida's favorite techniques.)

Problem A1. (see the following figure)

Given $d = 4, g = 3, f = 2$, and $a + b + c = 36$, let the circles shown have radius r. Show that $a = 16, b = 12, c = 8$.
    Using altitude k, Aida writes

$$\frac{r}{b} = \frac{f}{k}, \frac{r}{c} = \frac{g}{k} \iff \frac{bf}{r} = k, \frac{cg}{r} = k \quad \therefore bf - cg = 0.$$

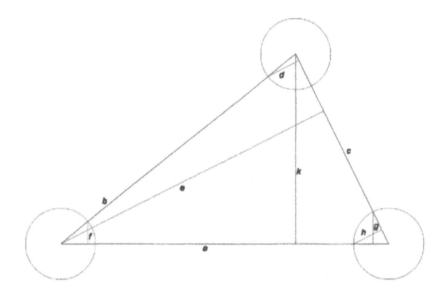

By analogy—Aida includes altitude e in the figure to give a hint of what he is up to, but not the third altitude—we have $bd - ag = 0$ and $cd - af = 0$, so $b = \frac{ag}{d}, c = \frac{af}{d}$. Hence

$$a + \frac{ag}{d} + \frac{af}{d} - 36 = 0$$

or

$$a(d + f + g) - 36d = 0, \qquad b(d + f + g) - 36g = 0,$$
$$c(d + f + g) - 36f = 0$$

$$\frac{36d}{d + f + g} = a, \qquad \frac{36g}{d + f + g} = b, \qquad \frac{36f}{d + f + g} = c$$

and the result follows.

In effect, Aida has proven the Law of Sines to solve the problem. But he doesn't say the radii of the circles are 1, which give segments d, f, g the value of the sines of the corresponding angles; on the contrary, each is greater than 1, whereas $|\sin x| \le 1$. Problem A1 represents the sort of exact solution Aida regards as ideal, but only because the lengths d, g, f, are given and all the r cancel out.

Problem A2. (same figure as before)

This time, Aida gives labels $\overarc{ab} = C = 15, \overarc{bc} = A = 24, \overarc{ac} = B = 20$, lets $a + b + c = 1{,}000$, and explicitly invokes what we would call the Law of Sines. The circumferences of the small circles are each 120; i.e., Aida is measuring arcs in units of 3°. Their sum should therefore be 60, not $A + B + C = 59$. Was this a slip of the brush or a mistake due to being unaccustomed to working with angles?

   If we correct Aida's $C = 15$ to $C = 16$, we get $a = 371.474, b = 338.261, c = 290.265$, which are close to Aida's calculated values $a = 376.557, b = 343.090, c = 280.132$. The discrepancies are probably due to the fact that Aida did not use trigonometric tables, as Honda assumed the reader would. Instead, he tries to calculate the sines directly from the data using what was known as *kohaijutsu* 'arc technique' 弧背術. Given a circle of diameter d with a chord a substending an arc of length s, *kohaijutsu* asserts that $a = \frac{s}{d}\left[1 - \frac{1}{3!}\left(\frac{s}{d}\right)^2 + \frac{1}{5!}\left(\frac{s}{d}\right)^4 - \cdots\right]$ (Satō et al. 2009: 502, slightly corrected). In modern terms, $\frac{s}{d}$ is the radian measure of half the arc and the chord is $2 \sin \frac{s}{d}$. Thus, letting $\frac{s}{d} = x$, *kohaijutsu* amounts to summing the first few terms of the Maclaurin series

$$\sin x = \sum_{k=0}^{\infty} \frac{(-1)^k}{(2k+1)!} x^{2k+1} = x - \frac{x^3}{3!} + \frac{x^5}{5!} - \cdots.$$

For three reasons, it is remarkable that this infinite series pops up in a book by Aida.

First, it originated in the Seki school of *wasan*, headed at the time by Fujita Sadasuke, with whom Aida had a long-running dispute. Arima Yoriyuki 有馬頼徸 (1714–1783), a *daimyō* who was taught by Yamaji Nushizumi 山路主住 (1704–1773) of the Seki school and excelled in mathematics,[28] revealed the secret in his *Shūki sanpō* of 1769, which, as noted on p. 22, Aida had read. It was said the Arima was upset by the amount of money Yamaji was charging him for lessons, but that is a story that sounds too good to be true.

Second, it is remarkable that the Japanese had discovered this infinite series since they lacked an explicit theory of differentiation or of the sine as a continuous function, upon which the usual derivation of the Maclaurin series for the sine depends.[29] Such infinite series were known to Mādhava (c. 1340–c. 1425) of the Kerala region (the southwesternmost coast) of India centuries before Newton, Gregory, Taylor, and Maclaurin used them,[30] and it is not inconceivable that Jesuit missionaries brought knowledge of them to Japan from the Portuguese State of India.[31] But as Seki (1642–

---

[28] Yoriyuki, the seventh lord of the Kurume domain in the southern part of what is now Fukuoka prefecture.

[29] The Maclaurin series is $f(x) = f(0) + f'(0)x + \frac{f''(0)x^2}{2!} + \frac{f'''(0)}{3!} + \cdots + \frac{f^{(n)}(0)}{n!} + \cdots$, where f is any real- or complex-valued function that is infinitely differentiable at 0. Because $\frac{d}{dx}\sin x = \cos x$, $\frac{d}{dx}\cos x = -\sin x$, $\sin 0 = 0$, and $\cos 0 = 1$, the infinite series for $\sin x$ follows at once.

[30] For more on Mādhava, see Van Brummelen 2009: 113 ff.

[31] The Portuguese State was founded in 1505; Francis Xavier arrived in Japan in 1549.

1708) was active after the Portuguese had been expelled from Japan, any connection with Indian mathematics, if one existed, would more likely have been by way of intermediary Chinese sources.[32]

Finally, we see that Aida's idea of angle measure (the value of x in the series) is similar to radian measure[33] except that, for Aida, who thought in terms of diameters rather than radii, the basic unit was what we would call two radians.

Leaving aside the question of how the Japanese discovered the infinite series for sin x, *Oranda sanpō* makes it clear that Aida was uncomfortable with the use of trigonometric tables. Kobayashi says he cannot help thinking of Aida as fundamentally different from earlier *wasanka* because, unsatisfied with merely explaining Dutch methods in traditional Japanese terms, he wanted to recast and integrate them into Japanese practice. Was this because he had a genuine horror of infinite series,[34] or because of his antipathy to the Seki school? Whatever the reasons may have been, Aida evidently did not share Honda's unabashed enthusiasm for Western techniques, which, if nothing else, shows us something about the intellectual diversity of mathematicians in Edo period Japan.

---

[32] On the possibility of a Kerala-China connection, see Joseph 2011: 434–35. But Kobayashi (p.c.) thinks a connection is highly doubtful, and Fukagawa and Rothman, who identify Takebe Katahiro (1664–1739), Seki's most illustrious disciple, as the first Japanese to use infinite series (2008: 303–4), state flatly that his methods "did not appear in Chinese mathematics" (75).

[33] An angle of one RADIAN subtends an arc of a circle of the same length as the circle's radius.

[34] One is reminded of the comment ascribed to Leopold Kronecker (1823–1891), who had similar aversions: *Die ganzen Zahlen hat der liebe Gott gemacht, alles andere ist Menschenwerk* [The good Lord made the integers; all else is human work].

Appendix B: Japanese Romanizations

Japanese scientists and mathematicians often use the Nippon-shiki style of romanization, or the slightly different official (and ISO standard) Kunrei-shiki system, rather than the Hepburn romanization, which I use here in keeping with American bibliographic practice. One often finds a jumble of romanizations in English-language literature (e.g., Fukagawa & Pedoe 1989, Fukagawa & Rothman 2008), including the occasional syllable form, such as *jyo*, that is not a proper spelling in any recognized system of romanization.

The basic syllables (technically MORAE) of Japanese are transcribed the same way in all three systems of romanization except for the following:

Current hiragana	Hepburn	Kunrei-shiki	Nippon-shiki
くゎぐゎ	ka, ga (*older* kwa, gwa)	ka, ga	kwa, gwa
し	shi	si	si
しゃしゅしょ	sha, shu, sho	sya, syu, syo	sya, syu, syo
じ	ji	zi	zi
じゃじゅじょ	ja, ju, jo	zya, zyu, zyo	zya, zyu, zyo
ち	chi	ti	ti
ちゃちゅちょ	cha, chu, cho	tya, tyu, tyo	tya, tyu, tyo
ぢ	ji	zi	di
ぢゃぢゅぢょ	ja, ju, jo	zya, zyu, zyo	dya, dyu, dyo
つ	tsu	tu	tu
づ	zu (*older* dzu)	zu	du
ふ	fu	hu	hu
を	o (*older* wo)	o	wo

"Long vowels" such as Hepburn ō and ū are ô and û in the other two systems. The string -*tt*- in words like *nittyû* in those systems are -*tch*- in Hepburn (*nitchū*). A few users of Hepburn continue to write the mora nasal phoneme *n* ん as *m* before *m*, *b*, and *p* rather than *n*, but this ceased being standard practice among librarians and lexicographers decades ago.

References

Alsina, C., and R. B. Nelsen. 2007. On the Diagonals of a Cyclic Quadrilateral, *Forum Geometricorum* 7: 147–49.

Bankoff, L. 1983. A Mixtilinear Adventure, *Crux Mathematicorum*, 9: 2–7.

Fukagawa, H., and D. Pedoe. 1989. *Japanese Temple Geometry Problems*. Winnipeg: Charles Babbage Research Centre.

————, and J. F. Rigby. 2002. *Traditional Japanese Mathematics Problems of the 18th and 19th Centuries*. Singapore: SCT Press.

————, and T. Rothman. 2008. *Sacred Mathematics: Japanese Temple Geometry*. Princeton: Princeton University Press.

Hirata, Kōichi. 2006. Sakuzu kyōzai to shite no sangaku [*Sangaku* as a teaching resource for geometric constructions], 12th History of Mathematics Research Presentation Meeting (23 October).
Link at http://www.ed.ehime-u.ac.jp/~hirata/publish/051023-hirata.pdf.

Horiuchi, Annick. 2002. Honda Toshiaki (1743–1820) ou l'Occident comme utopie. In *Repenser l'ordre, repenser l'héritage: paysage intellectuel du Japon (XVIIe–XIXe siècles)*, by Frédéric Girard, Annick Horiuchi, and Mieko Mace, pp. 411–48. Geneva: Droz.

Keene, Donald. 1969 [1952]. *The Japanese Discovery of Europe, 1720–1830*. Revised and expanded edition. Stanford: Stanford University Press.

Knobloch, Eberhard, Hikosaburō Komatsu, Dun Liu, eds. 2013. *Seki, Founder of Modern Mathematics in Japan: A Commemoration on His Tercentenary*. Springer Proceedings in Mathematics & Statistics, vol. 39.

Kobayashi, Tatsuhiko. 1993. Aida Yasuaki-cho *Oranada Sanpō* sairon 会田安明著『阿蘭陀算法』再論 [A reconsideration of *Oranda sanpō* by Aida Yasuaki]. *Sūgaku-shi kenkyū* 数学史研究 [Studies in the History of Mathematics], no. 138 pp. 1–12.
Link at http://www.wasan.jp/math_indexj.html.

Kornicki, Peter F. 1998. *The Book in Japan : A Cultural History from the Beginnings to the Nineteenth Century*. Leiden; Boston: Brill.

Joseph, George Gheverghese. 2011. *The Crest of the Peacock: Non-European Roots of Mathematics.* 3rd ed. Princeton: Princeton University Press.

Nishida, Tomomi. 2013. Manuscripts in the Edo Period: Preliminary Study on Manuscripts Written by Seki Takakazu. In Knobloch, Komatsu, and Liu 2013, pp. 353–55.

Protasov, V. 1992. Vokrug Teoremy Fejerbaxa, *Kvant,* 23: (9) 51–58.

———. 1999. The Feuerbach Theorem, *Quantum,* 10: (2) 4–9 .

Rabinowitz, Stanley. 2006. Pseudo-Incircles. *Forum Geometricorum* 6: 107–15.

Rubinger, Richard. 1982. *Private Academies of Tokugawa Japan.* Princeton: Princeton University Press.

Satō, Ken'ichi, et al. 2009. *Wasan no jiten* [Handbook of *wasan*]. Tōkyō: Asakura shoten.

Smith, David Eugene, and Yoshio Mikami. 1914. *A History of Japanese Mathematics.* Chicago: Open Court.

Stephan, John J. 1983. Honda Toshiaki. In *Kōdansha Encyclopedia of Japan,* 3: 219–20.

Unger, J. Marshall. 2010. A New Proof of a "Hard but Important" *Sangaku* Problem, *Forum Geometricorum,* 10: 7–13. Edited with additional commentary, problems 16 and 17 at http://u.osu.edu/unger.26/files/2014/04/Sangaku-12zn2jo.pdf.

Van Brummelen, Glen. 2009. *The Mathematics of the Heavens and the Earth: The Early History of Trigonometry.* Princeton: Princeton University Press.

———. 2013. *Heavenly Mathematics: The Forgotten Art of Spherical Trigonometry.* Princeton: Princeton University Press.

Index

CORNELL EAST ASIA SERIES

CORNELL
East Asia Series

eap.einaudi.cornell.edu/publications

Milton Keynes UK
Ingram Content Group UK Ltd.
UKHW010653280723
425913UK00002B/4